THE REFORM OF THE COMMON AGRICULTURAL POLICY

The Case of the MacSharry Reforms

THE REFORM OF THE COMMON AGRICULTURAL POLICY
The Case of the MacSharry Reforms

ADRIAN KAY

Government Economist on the European Fast Stream, UK

CABI *Publishing*

CABI *Publishing*
CAB INTERNATIONAL
Wallingford
Oxon OX10 8DE
UK
Tel: +44 (0)1491 832111
Fax: +44 (0)1491 833508
Email: cabi@cabi.org

CABI *Publishing*
Suite 3203, 10 E. 40th Street
New York, NY 10016
USA
Tel: +1 212 481 7018
Fax: +1 212 686 7993
Email: cabi-nao@cabi.org

A catalogue record for this book is available from the British Library, London, UK.

Library of Congress Cataloging-in-Publication Data
Kay, Adrian.
The reform of the common agricultural policy : the case of the MacSharry reforms / Adrian Kay.
p. cm.
Includes bibliographical references and index.
ISBN 0-85199-298-6 (alk. paper)
1. Agriculture and state--European Union countries. I. Title.
HD1918.K393 1998
338.1'84--dc21 98-24191
CIP

ISBN 0 85199 298 6

Typeset by Wyvern 21, Bristol.
Printed and bound by Biddles Ltd, Guildford and King's Lynn.

Contents

Abbreviations

AMS	aggregate measure of support
BSE	bovine spongiform encephalopathy
BTE	budget and trade effect
CAP	Common Agricultural Policy
CoAM	Council of Agriculture Ministers
COPA	Comité des Organisations Professionnelles Agricoles
COREPER	Committee of Permanent Representatives
CP	compensatory payments
CRL	co-responsibility levy
DBV	Deutsche Bauernverbrand (German Farmers' Union)
DIP	direct income payment
DM	Deutschmark
EAGGF	European Agricultural Guarantee and Guidance Fund
ECSC	European Coal and Steel Community
ECU	European Currency Unit
EEC	European Economic Community
EEP	Export Enhancement Program
EIPA	European Institute of Public Administration
EMS	European monetary system
EP	European Parliament
ERM	exchange rate mechanism
EU	European Union
FT	*Financial Times*
GATT	General Agreement on Tariffs and Trade
LUFPIG	Land Use and Food Policy Inter-Group in the European Parliament
MAE	modern agricultural enterprise (as defined in the Mansholt Plan)
MAFF	Ministry of Agriculture, Fisheries and Food (UK)
MCA	monetary compensation amount
MGA	maximum guaranteed area

MGQ	maximum guaranteed quantity
Mha	million hectares
NFU	National Farmers' Union (UK)
NTB	non-tariff barrier
OECD	Organization for Economic Co-operation and Development
PE	partial equilibrium
PPF	political (or policy) preference function
PSE	producer subsidy equivalent
PU	production unit (as defined in the Mansholt Plan)
SAFER	société d'aménagement foncier et d'établissement rural
SAP	set-aside payments
SCA	Special Committee on Agriculture
SMU	support measurement unit
TE	tariff equivalent
UA	Unit of Account
VES	variable export subsidy
VIL	variable import levy
WTO	World Trade Organization

Chapter 1

Introduction

1.1 BACKGROUND

From the point of view of a student of economics, agriculture is an industry which holds much interest. Almost throughout the developed world it is subject to government intervention. These interventions are grouped under the heading 'agricultural policies'. The longevity and scale of agricultural policies in advanced industrial economies have taken agricultural markets well away from the textbook micro-economics description of a perfectly competitive market, and agriculture is one of the most heavily regulated industries in advanced industrial economies. For this reason, any account of agricultural policy must include some analysis of the role of government.

The Common Agricultural Policy (CAP) is the agricultural policy of the European Union (EU).[1] It has existed, at least in some form, since the Stresa conference of the original six members of the EU in 1958. It was the first common policy of the EU and has remained its largest in terms of share of the EU budget, accounting for around 47.5% of budget expenditure in 1994. On these terms alone the CAP demands attention. In addition, its constituency has been shrinking. The agricultural sector accounted for almost 14% of total EU employees in 1970, but this had fallen to under 6% by 1992. It has appeared resistant to a substantial reform to tackle either the policy's failure to deal with the farm income problem or the surpluses and budget expense that its operation has generated. Thus the CAP, and more particularly reforms of the CAP, are of interest to a student of government.

The MacSharry reforms, enacted in May 1992, are the latest section in the history of the CAP, following reforms in 1984 and 1988. They were different from previous reforms of the CAP in a number of ways: (i) in the circumstances in which they were enacted, there was not the familiar sense of EU crisis in which the European Council was required to intervene in the CAP decision-making system[2] in order to agree a response to pressing circumstances; (ii) the factors which prompted the reform process appear to have been more varied than in previous reforms; (iii) the type of CAP that the MacSharry reforms have brought about is new, even though the overall effects of its operations may remain similar.

The main objective of this book is to provide an analysis of why the MacSharry reforms were enacted, when they were enacted and the way that they were enacted. The first question refers to the pressures which started and finished the MacSharry reform process. An answer to the second question requires an account of the context of those pressures; the circumstances which affected the timing of the reforms. The third question looks at the substance of the reforms; why that type of reform was enacted in response to the pressures and circumstances outlined for the answers to the first two questions. In order to provide such an account, this book has adopted an inter-disciplinary approach; that is, it examines alternative approaches to the standard neo-classical agricultural economics analyses of the CAP. This is not an economics book. Although the work is relevant to economists, only Chapter 2 has a significant agricultural economics component. The inter-disciplinary approach adopted recognizes that agricultural policy is made by governments and is affected by economic and political pressures as well as having economic and political effects. This work will be of interest to a range of CAP analysts outside the strict agricultural economics domain.

1.2 THE CAP DECISION-MAKING SYSTEM

Since agricultural policy is made by governments, it is the product of some decision-making system. The CAP decision-making system consists of three elements: the first is the pressures or inputs in the system; these prompt the second element, the process which translates those pressures or inputs into the third element of the system – the actual observed decision about the CAP (the system's output).

Figure 1.1 sets out these elements. Element 1 consists of pressure groups, national governments and objective economic circumstances. All of these feed into the policy process (element 2). It is in the policy process that the first elements of the system are translated into actual observed policy decisions. It is titled here the *black box* of CAP decision-making. The term is used to highlight the common assumption, in the political economy literature, that the numerous and relatively complex causal links which operate in the policy process are unimportant in the output of any decision-making system. Element 2 is, therefore, assumed away as a black box. This book directly challenges that assumption. It aims to shed light on the black box and demonstrate that the policy process is the most important factor in understanding the CAP decision-making system.

Section 1.4 describes the black box of the decision-making system in

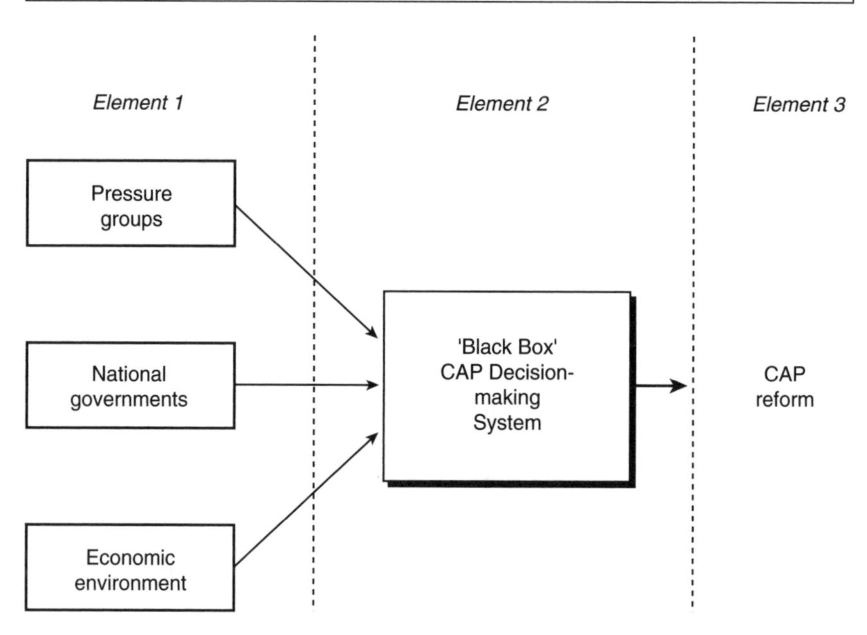

Fig. 1.1. The CAP decision-making system.

a formal way, in terms of the prescribed competencies of each of the institutions involved. The term 'black box' also helps to define a research agenda into which this book aims to fit: how inputs into and pressures on the decision-making system are translated into observed public policies. This book is a study of a policy process. It starts with the third element of the CAP decision-making system, the MacSharry reforms, and works back to the questions of why the reforms were enacted, when they were and the way that they were.

Chapter 5 sets out three competing analytical frameworks of CAP reforms for application to the MacSharry reforms. Each framework focuses on different elements in the CAP decision-making system as the most important in understanding the output of that system; that is, the decision to reform the CAP. The chapter also selects one of the three frameworks as appropriate for understanding the MacSharry reforms.

1.3 ELEMENT 1

1.3.1 Pressure Groups

There are many and varied organized interest groups in the CAP decision-making system. The national farm unions have traditionally been regarded as having strong influence over their national representatives in the Council of Agriculture Ministers (CoAM). The Comité des Organisations

Professionnelles Agricoles (COPA) is a supranational umbrella organization which brings each of the national farm unions together to agree a common position with regard to Commission proposals for the CAP. Other interested groups with respect to the CAP are consumer groups, environmental groups and agribusiness organizations. The influence of agribusiness organizations has at times been seen as important, but in the absence of a public forum or union its extent is difficult to assess. So far as environmental groups are concerned, it has been argued that they have played an increasing role in national and EU agricultural policy agenda since the mid-1980s. Consumer groups have traditionally been regarded as ineffectual, but the MacSharry reforms may provide substance to the argument that their influence has grown in the 1990s.

1.3.2 National Governments

National governments have specific functions in the black box of the CAP decision-making system. How they exercise those functions is determined by factors outside that system, domestic politics being an important variable in their CAP decisions.

Although subnational or regional governments often have relationships with various parts of the Commission, their input into CAP reform comes only through representations to their national government. They do not have a formal position in CoAM negotiations.

1.3.3 Economic Environment

EU agricultural markets are subject to the vagaries of nature as well as the vagaries of government regulation. Further, there is a world market for agricultural products. The EU does not sit in subsidized isolation. A combination of the state of both these markets creates economic circumstances for EU farmers and the EU budget. These are inputs into the black box part of the CAP decision-making system.

1.4 ELEMENT 2

This section presents a description of the institutions included in the black box of CAP decision-making (Fig. 1.2). It is deliberately formulaic, outlining the role of each institution in the policy process. This gives only a limited insight into the factors which are important in a CAP reform process. The description is included here for two reasons. First, the institutions involved in CAP decision-making are introduced at various points in this book as crucial factors in understanding CAP reforms and the MacSharry reforms in particular. Second, in a book about EU public policy, it is important to understand the process through which policies are designed and implemented.

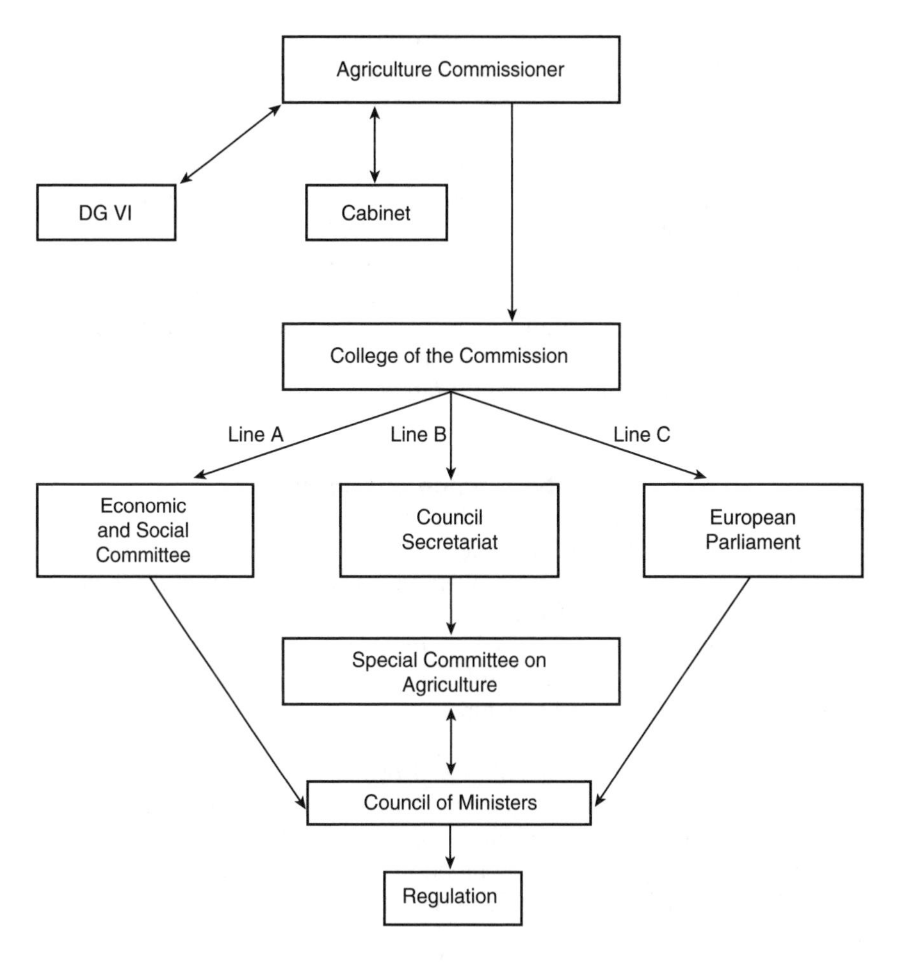

Fig. 1.2. The black box element of the CAP decision-making system.

1.4.1 The Commission

The Commission enjoys the sole right of initiation of legislation in the EU. The Council cannot take a decision on a subject without a proposal from the Commission. It may agree something different from the Commission's proposal but this requires unanimity among member states in the Council. As Meester and Van der Zee (1993, p. 134) state, 'The Commission's proposal is formally and also materially the basis of the subsequent negotiations.' The main day-to-day work of the Commission is the management and implementation of policies legislated in the Council. However, proposals for the reform of the CAP are generated in the Commission; hence it is an important institution in a CAP reform process.

The College sits at the head of the Commission. It consists of 20 commissioners (17 until the accessions of Austria, Finland and Sweden to the EU in 1995): two from each of the large member states and one from each of the remaining member states. Commissioners are political appointees nominated and appointed by the agreement of the member state governments. They are nominally independent, but the research for this book confirmed a tendency to favour the interests of the government of their country of origin (often with strong lobbying from that government).

Each of the commissioners is allocated an area of responsibility under the Treaty of Rome. They are assisted by their cabinet – a small staff of advisers who act as the commissioner's think-tank. They also act as a line of communication with the Commission Services, the 23 Directorates-General (DGs) which are the administrative departments of the EU. DG VI is in charge of agricultural policy.

1.4.2 The Council of Ministers

The Council of Ministers is the primary legislative body of the EU. There is a separate Council for each policy area, each being composed of representatives of the member state governments. The Council of Agricultural Ministers (CoAM) meets at least once a month (along with the Foreign Affairs Council, it is the Council which meets most frequently). It holds the ultimate power to enact a CAP reform. In addition it agrees the various institutional price levels and structural policy, and has some influence over trade policy.

The European Council is composed of the heads of state or government of the member states together with the Commission President and meets at least twice a year. Because of its political power, it has sometimes resolved serious issues of contention in the development of the CAP, as we shall see in Chapter 3 which presents the histories of the reforms of 1984 and 1988.

Each separate Council is chaired by a representative of one of the member states; member states hold the presidency on a 6-monthly rotation. The presidency acts as arbitrator and mediator in the Council and can engineer compromise initiatives. This role should be considered in conjunction with that of the Agricultural Commissioner. The balance of power between these two individuals is variable, depending on a number of different circumstances.

The Council is assisted in its deliberation by a secretariat. This set of officials is independent of the Commission and the member states. It provides documentation, advice and background material for the committees which process the draft legislation before formal discussion and negotiation by the Council.

Most CAP legislation is agreed by qualified majority voting in the CoAM. Each member state has a weighted vote in the Council, roughly according to its population size. The weights run from ten votes (Germany, France, Spain, Italy and the UK) down to one vote (Luxembourg). A qualified majority prior to the 1995 accessions was 54 out of 76 votes. Since the 1995 accessions, it has been 62 votes out of 87. If the Council wants to agree a policy different from that proposed by the Commission, unanimity is required. In practice, the Commission will usually accept the final presidency compromise as its final proposal if it appears that that proposal commands a qualified majority in the Council.

1.4.3 The European Parliament

The powers of the European Parliament (EP) have remained marginal in agricultural policy decisions, despite growing significantly in certain areas of EU legislation after the Single European Act (1986) and the Treaty on European Union (1991). For Council decisions about the CAP, the EP only has to be consulted; the CoAM can, and often does, ignore its formal opinion. Further, the budgetary powers of the EP, which are extensive for some types of EU expenditure, are limited for CAP expenditure. Spending through the European Agricultural Guarantee and Guidance Fund (EAGGF) is classified as *compulsory* expenditure. This means that the EP has the right only to propose modifications to the draft budget; it cannot reject it. These can be added to the draft budget only by the agreement of a qualified majority in the Council. Failure to achieve a qualified majority in the Council means that the modifications are not made. This is significantly less influence than the EP enjoys for *non-compulsory* expenditure.

In the case of the MacSharry reforms, the Farm Committee of the EP failed to agree on a report. Such a report usually forms the basis of the EP's opinion on proposed CAP legislation. In the case of the MacSharry reforms no substantive opinion was agreed by the EP and therefore the influence of the EP was marginal in the reform process.

1.4.4 The Economic and Social Committee

The need to 'have regard for the opinion of the Economic and Social Committee' is established in the Treaty of Rome (Article 198). The Committee has little influence on policy-making. It is a body of 189 members drawn from various interest groups. Its opinion on the agricultural reform proposals in COM (91) 258 (Commission of the European Communities, 1991b – the MacSharry reform proposals) was delivered to the CoAM on 26 February 1992 and published in the *Official Journal* on 21 April 1992 (OJ C 98). The opinion rejected the price reductions proposed and the regional yield factor in the compensatory payments. Neither of these criticisms was addressed by the Commission or the CoAM.

1.4.5 Special Committee on Agriculture

The role of the Special Committee on Agriculture (SCA) is hard to assess. It is composed of representatives of the member states' governments, specifically senior civil servants, advisers and experts. The body never formally votes but can develop agreement on points in the Commission's proposals (known as *A points*) which can be sent to the CoAM for confirmation. Further, the SCA provides technical analysis of proposals or scenarios which may become the subject of political debate in the Council. The role of the SCA in the MacSharry reforms is hard to assess as its findings and proceedings have never been made public. The SCA is 'special' because in all other policy areas this function is performed by the Committee of Permanent Representatives (COREPER).

1.5 ELEMENT 3

Decisions about the CAP, including CAP reforms, are the output of the system described above. A complementary objective of this work, associated with the central objective of understanding the MacSharry reforms, is to develop a richer understanding of the CAP decision-making system. There are three basic questions which confront analysts of the CAP as the product of a decision-making system: (i) in what sense is the CAP stable, and how is this property explained?; (ii) at the same time, why is it subject, periodically, to reform?; (iii) what explains the nature of those reforms?

Stability in the CAP can be defined at two levels. First, the CAP can be thought of as evolving over the long term through conflict and responses to that conflict. The operation of the CAP and its associated effects (financial, economic, political, international, etc.) comes into conflict with interests within the EU and external to the EU. This conflict produces pressure for change. However, the response of the CAP decision-making system to such pressure in the short run is muted. Examining the CAP on a year-by-year basis, its development seems to be dominated by the status quo; any change takes place on an incremental basis and the policy seems resistant to reform. These characteristics of the CAP decision-making system have led to the CAP being described as stable. When reforms are, periodically, enacted, they never fully satisfy the pressures for change and contain within them the catalysts of pressures for further change. This leads to a second and deeper understanding of the stability of the CAP.

The distribution of the costs and benefits of the CAP at a national aggregate level has been relatively stable over the history of the CAP. Changes in world markets, the structure of agriculture and changes in CAP regimes have altered the distribution of costs and benefits between different types and sizes of farms, and between consumers and taxpayers, yet each member state's net pay-off from the CAP has remained relatively stable (Ackrill *et al.*, 1995). The CoAM has consistently supported a fixed political bargain between the member states. Chapter 7, on the effects of the MacSharry reforms, surveys the evidence in this regard.

In this context, CAP reforms can be regarded as an outbreak of turbulence against a long-run trend of stability. Turbulence refers to the political battles associated with reforms of the CAP. The system of distributing CAP benefits is changed, the incentive structure for farmers is reformed but the underlying pattern of national member states' net pay-offs from the CAP remains stable. A CAP reform which upset this balance of national member states' net pay-off from the CAP would be something more than turbulence. Such a reform could be termed *radical.*

A consideration of feasible policy options is useful in elucidating the distinction between turbulence and stability. There have been a number of different systems in the history of the CAP for distributing the benefits of agricultural intervention. The CAP has been most notably reformed in 1984, 1988 and 1992. Each of these reformed systems of the CAP was a feasible policy option, in the sense that they supported the second level of stability outlined above. Each system of the CAP has been stable at the first two levels, but not unique. The movement within the set of feasible policy options is associated with turbulence. Turbulence is the word chosen to describe the reform process; it is intended to capture the observation of political battles, machinations, and the 'blood and sweat' associated with moving the CAP from one non-unique feasible policy option to another.

The demand to provide an answer to the three questions set out at the beginning of this section raises issues about the order of explanation. The standard way is to start with an explanation of the stability of the CAP. From this starting-point, an account of CAP reform and turbulence can be developed. However, this book holds that the order of explanation should start with the reform process. The CAP reforms of 1984, 1988 and 1992 are the salient features of the history of the CAP since 1980. By explaining the nature of turbulence and why it does not affect the underlying stability of the CAP, the characteristic of stability is explained. It is easier to explain stability through an account of turbulence than turbulence through an account of stability.

The order of explanation in this piece of work starts from an account of the MacSharry reforms. From this particular episode, a more general understanding of the CAP reform process can be developed using the frameworks outlined in Chapter 5. This aids a richer understanding of the CAP decision-making system and the history of the CAP.

1.6 LAYOUT OF THE BOOK

The book consists of two parts. Part I (Chapters 2–4) sets up a framework for analysing the MacSharry reforms of May 1992. The starting-point is a consideration of the limitations of a strictly agricultural economics approach. Chapter 2 considers the neo-classical agricultural economics critique of the CAP; that is, the standard welfare analysis of the CAP as a means of transferring income to farmers plus a consideration of the consequences for the EU budget of the trends in EU production under the CAP. Chapter 3 describes the history of the reform of the CAP and demonstrates the limited influence of the CAP reform prescriptions of neo-classical agricultural economics on actual, observed CAP reforms. Chapter 4 describes the possible causes of the reform process by outlining the emerging pressures, external and internal, on the CAP in the early 1990s.

Part II of the book (Chapters 5–8) is the case study of the MacSharry reforms. Chapter 5 introduces the public choice paradigm as an alternative to neo-classical agricultural economics as a means of understanding CAP reforms. As is stated in Section 1.2, it develops, within the public choice paradigm, three rival frameworks for understanding CAP reforms: the interest groups, the prominent players and the institutions frameworks. Each of these focuses on different elements in the CAP decision-making system as influential in the outcome of a CAP reform process. Section 5.9 selects the institutions framework as the most suitable to employ in the construction of the MacSharry case study.

The policy process which links the cause and effect of the MacSharry reforms is the subject of Chapter 6. This is the key chapter of Part II. It examines the causal links in the chain between the pressures for reform (Chapter 4) and the effect of the reforms (Chapter 7), and gives an account of how the black box is operated in the MacSharry reform process. Chapter 8 considers whether the evidence of the MacSharry reform process might be better interpreted using one of the rival analytical frameworks of CAP reforms rejected in Chapter 5. Chapter 9, the conclusion, draws together the two main strands of the book, an understanding of the MacSharry reform process and how that understanding contributes to analyses of how the CAP decision-making system operates.

NOTES

1. The European Union (EU) has, at various times in its history, been called the European Economic Community (EEC) and the European Community (EC). For convenience, this book assumes that the European Union has been the EU throughout its history. Only where it has been necessary to make a distinction for reasons of describing history are the terms EC and EEC used in the book.

2. Sartori (1976, p. 46), when discussing party systems, defined a system as having two properties: '(i) the system displays properties that do not belong to a separate consideration of its component elements and (ii) the system results from, and consists of, the patterned interaction of its component parts, thereby implying that such interactions provide the boundaries, or at least the boundedness of the system'. Such a definition covers the use of 'decision-making system' or 'political system' in this book.

Chapter 2

An Economic Analysis of the Common Agricultural Policy

This chapter has three aims. The first is to provide a description of the workings of the CAP and the different instruments employed to achieve the objectives for a common agricultural policy laid down in the Treaty of Rome. This will introduce the terminology which will be used through the rest of the book.

The second aim of this chapter is to provide a basic economic critique of the CAP. Its costs and benefits will be described from the perspective of standard, neo-classical welfare economics. The benefits will be judged against the objectives for the CAP stated in Article 39 of the Treaty of Rome. The costs comprise the direct financial consequences of the CAP as well as the indirect economic costs. Taken together, the first two aims of this chapter will provide an agricultural economics interpretation of the CAP and its evolution. The analysis is concerned with demonstrating the inefficiency of the policy in terms of achieving its declared aims.

The third aim is to outline the academic response to the diagnosis of the CAP. The main CAP reform proposals made by agricultural economists during the history of the CAP will be outlined.

This chapter is organized into three separate sections, corresponding to the aims set out above. Section 2.1 looks at the general economic rationale for government intervention in agriculture. Section 2.2 sets out the basic economics of the development of the CAP and the emergence of pressures for its reform. Sections 2.3.1 and 2.3.2 present the standard partial equilibrium welfare analysis of a tariff and an export subsidy. This provides the diagnosis of the CAP as an inefficient means of redistributing income to farmers. Section 2.3.3 surveys the prescriptions offered by agricultural economists to this diagnosis.

2.1 ECONOMIC JUSTIFICATION FOR AN AGRICULTURE POLICY

A strict agricultural economics approach to the analysis of the CAP starts with a justification of government intervention in agriculture. The rationale for intervention centres on the special economic characteristics of agriculture – specifically, the existence of a chronic farm income problem. The factors which contribute to the farm income problem can be categorized according to whether they affect the demand side or the supply side of agricultural markets.

The demand side of an agricultural market is characterized by Engel's law, which states that the income elasticity of the demand for food decreases as income increases. As European households' incomes are relatively high, their income elasticities of demand for food are low. It should be noted, however, that the income elasticities for certain foods have risen with incomes because of the quality, luxury or convenience aspects of the demand for food. The marketing margin in the food processing and retail industries has increased in certain areas of the food industry. Engel's law can be restated in terms of farm output; the demand for farm output has remained relatively stable as household incomes have risen in developed countries. However stated, low income elasticities of demand for agricultural products are a key part of the agricultural industry's relative decline as the Western European economy has developed. The other important aspect which has constrained the rate of growth of demand for agricultural products in Europe has been the low population growth which has characterized most of the European economies since the 1960s.

The limited growth in demand for agricultural products poses problems for the profitability and economic viability of the sector only when considered with the issue of supply in agricultural markets. It is here where agricultural economists' analysis of the farm income problem has been concentrated.

Supply in agricultural markets is subject to the problem of long lead times from production decisions to actual output. Price signals influence production decisions. An increase in price, signalling strong demand relative to supply, only finds an output response some period later when the original market conditions may have changed (in that commodity market or its close substitutes). This leads to a tendency for commodity markets to be over- or under-supplied at any time; price adjustments tend to be exaggerated. Cohen (1959) provides an analysis of these cycles in agricultural markets. This is one element of what farmers call the 'farm income problem'; that is, instability or variability in farmers' incomes due to exaggerated volatility in market prices.

Tweeten (1971), examining the supply of agricultural products, defines the farm income problem in terms of low rates of return on resources employed in the farm sector. In other words, farm resources earn less than their opportunity costs. This implies a disequilibrium in the resources employed by a farm: the mix of inputs is not optimal and this inefficiency implies low rates of return in the agricultural sector. Economic growth and inflation force constant adjustments in the efficient mix of farm resources just as they do in the non-farm sector. A characteristic of agriculture since the Second World War has been the widespread and consistent application of technological advancements. For a given level of inputs in the agricultural sectors of the industrialized world, there has been constantly expanding output. Demand for agricultural products, as described above, has neither rapidly increased nor exhibited a high price elasticity. The fact of output increasing quicker than demand has led to depressed prices and incomes for farmers.

The reasons why the agricultural sector has not seen an adjustment at the required rate in the resources employed (in order to avoid the situation described above) are the reasons why there is a farm income problem. The fixed asset theory is the main explanation of this phenomenon used by agricultural economists (Martin, 1958; Johnson, 1960; McCrone, 1962; Hathaway, 1963; Tweeten, 1971). The theory holds that farm resources, especially labour, tend to be relatively fixed in agriculture. They are not easily transferable to the non-farm sector. Johnson (1960) notes that a consequence of this resource immobility is that there typically exists a substantial divergence between the acquisition costs and the salvage value of farm assets.

Martin (1958) attributes the poor inter-sectoral labour mobility from agriculture, to the age structure of farmers, their poor education (which has become less of an issue since the 1950s) and lack of transferable skills. Martin (1958) broadens the debate about the limited inter-sectoral mobility of farm labour, pointing out that a farming life brings a number of intangible benefits, which lead to the acceptance of pecuniary incomes lower than elsewhere in the economy; farmers accept negative economic rent. Tweeten (1971) comments, also, that the opportunity cost of being in agriculture for a farmer or farm labourer is often zero or the level of unemployment benefit.

Howarth (1985), at the time of mounting EU surpluses, noted that 'It is farmers, not milk and wheat, which are in oversupply' (p. 51). The fact that farmers cannot or will not leave agriculture depresses agricultural incomes relative to other sectors in the economy. Low incomes have been used as a rationale for government intervention in the agricultural sector. Some kind of state support is almost universal in advanced, industrialized economies.

2.2 THE EVOLUTION OF THE CAP

This section makes the step from a general economic rationale of government intervention in agriculture to the specifics of the history of the CAP. The first important point to appreciate is that each of the original member states prior to the creation of the European Economic Community (EEC) had long-standing national agricultural support policies, as did the UK, Denmark and Ireland when they joined in 1973. Greece, Spain and Portugal had more limited state intervention schemes in place when they joined the Community.[1]

To understand how and why the CAP came into existence and became the EU's first common policy requires a review of diplomatic history.

2.2.1 The Establishment of the Common Agricultural Policy

The Spaak Committee Report commissioned by the 1955 Messina conference of the European Coal and Steel Community (ECSC), the forerunner to the EEC, laid a plan for a common market in the ECSC area and explicitly included agriculture. France, Italy and the Netherlands insisted that agriculture must be included in any common market that was envisaged in the drafting stage of the Treaty of Rome. Indeed, many have understood the EEC as an alliance between German industry and French agriculture. In addition, the importance of the German position in CAP reform negotiations will be shown in Chapters 3 and 6.

Article 39 of the Treaty of Rome (1957) sets out vague objectives for a common agricultural policy. These were put into more concrete form by the first Mansholt Plan, begun after the Stresa conference of 1958. The guiding principles of the CAP gradually became market unity, Community preference and financial solidarity. In 1962 the CoAM passed a timetable for common intervention prices for the domestic market, as well as threshold prices for the operation of variable import levies. Agriculture had little or nothing to do with the common market in industrial goods, because it was heavily supported by the Six prior to the CAP. The very fact that the acronym was CAP and not CAM (Common Agricultural Market) suggests that the debate was centred on a common intervention policy. Tracy (1994) provides a comprehensive account of the Stresa conference and the various interests and demands that were being negotiated there.

The CAP was the first European policy and remains the most extensive. Agriculture was probably the easiest area in which to start because there existed six national policies already; the abnegation of national responsibility for a difficult and expensive problem held some political attractions. This is known as the 'restaurant bill' phenomenon in welfare economics. When the bill is being paid for collectively there exists an incentive for individuals to spend more money than they would have done if it was the individual alone who was paying the bill. As the rest of this chapter will show, the financing of this restaurant bill has been one of the most controversial aspects of the CAP.

Agricultural economists have analysed the history of the CAP in terms of the objectives stated for a common agricultural policy in Article 39 of the Treaty of Rome:

1. To increase agricultural productivity by promoting technical progress and by ensuring the rational development of agricultural production and the optimum utilization of the factors of production, in particular labour;
2. Thus to ensure a fair standard of living for the agricultural community, in particular by increasing the individual earnings of persons engaged in agriculture;
3. To stabilize markets;
4. To ensure the availability of supplies; and
5. To ensure that supplies reach consumers at reasonable prices.

Article 39 is a statement of objectives which began to be interpreted and implemented at the Stresa conference of 1958. The three pillars of the CAP emerged in the following 5 years:

1. A single market, allowing the free movement of agricultural commodities;
2. Community preference, protecting EEC farmers from cheaper imports; and
3. Financial cohesion, reinforcing the *common* nature of both the policy and its shared funding through the EEC budget.

2.2.2 The CAP in Practice

The CAP as it evolved from Article 39 was a price support system; in practice this has meant direct intervention, or the threat of direct intervention, in EU agricultural markets to support producer prices. The central feature of this kind of price support system is that the burden of support falls on consumers rather than taxpayers (although the MacSharry reforms have shifted the burden of the support of European agriculture more towards taxpayers).

The market price of the EU was generally set at a level above world prices. The validation and support of this market price required the EU to administer three prices, as represented in Fig. 2.1.

In Fig. 2.1, domestic demand for an agricultural commodity is D_d and its domestic supply is S_d. The target price, P_t, represents the desired price in the EU agreed by the CoAM each year. To ensure that the market price exists within some range of the target price two further prices are administered. The threshold price, P_{th}, is the minimum price for imports entering the EU. It is below the target price in order to reflect some notional cost of transport from this port of entry to final market. A variable import levy (VIL) was set as the difference between the threshold price and the lowest offer price from exporters looking to sell their cereals in the EU (this price is a proxy for the world price, marked P_w on the diagram). The VIL is variable because although the threshold price is fixed, the world price moves with changes in world supply and demand.

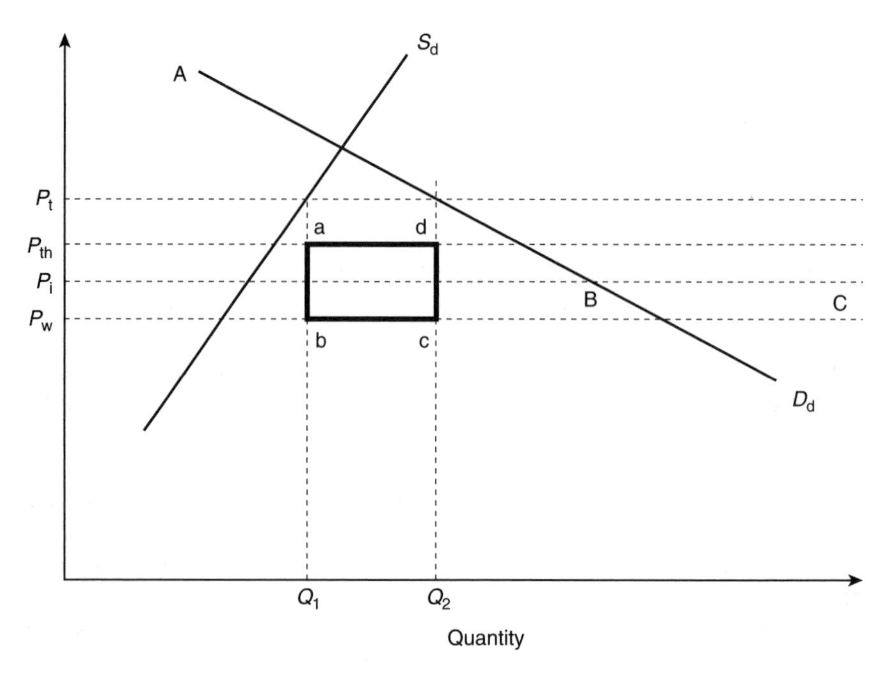

Fig. 2.1. The EU as a net importer of agricultural commodities.

The third administered price is the intervention price, P_i. This is the floor to the market, the price at which the EU, through a system of intervention buying using national governments, agrees to buy everything that farmers can supply at P_i. The effective demand curve for agricultural products is ABC.

Exports of CAP products from the Community receive a subsidy or refund. This means that the sale price received is the high internal EU price and not the lower world price. The variable export subsidy (VES) is equal to the difference between the market price in the EU and the price at which exporters sell in third-country markets.

A strict agricultural economics account of the development of the CAP is a history of how this system has increasingly been strained by a series of different pressures. It is these strains which have led to CAP reforms. This support system was designed for the agricultural situation in the six-member EEC (EEC6) in the 1960s. A common market with a common price in the EEC was effectively completed in 1969, a system of price support being seen as vital to the prosperity and continued development of European agriculture. It is generally considered that agreeing a common price increased average EU prices. The burden of price support could be placed on consumers; high prices were a feature of the national agricultural support systems of the EEC6, which relied on a continued wartime sentiment that agriculture was a necessary industry. Most pertinent for the subsequent development of the CAP was that the budget implications at the time were favourable (indeed, the CAP was originally intended to be self-financing). As a net importer of agricultural commodities, the EEC budget gained more in VILs than it paid out in VESs. This is represented in Fig. 2.1 by the box abcd.

The link between the budget cost of the CAP and the excess of production over consumption came to dominate agricultural economists' thinking about the problems of the CAP in the 1980s. This complemented an existing critique of the CAP in terms of its ineffectiveness at supporting low-income farmers (see Section 2.3). The growth in EU productivity and output is well documented. Productivity gains have been achieved in all agricultural sectors, and have fed through to an increase in total production. All agricultural economics models involve some guess at this trend of rising productivity. However, the productivity growth has not solved the farm income problem. Cochrane (1958) posited the theory of an agricultural treadmill. The theory starts from the point that innovations which increase supply will tend to depress prices as there is a low price elasticity of demand in advanced countries. Supply-boosting innovations are generally introduced by a small number of pioneers who take the risk with new techniques, processes and machinery. As the innovations increase aggregate supply, the market price falls, and producers using the old technique (and therefore producing at a higher cost) begin to incur a loss. The rate of adoption increases, aggregate supply increases further and the market price falls further. All farmers are eventually forced to adopt the new technique to minimize their loss of income, given a falling market price.

Figure 2.2 shows that productivity levels and total production levels in European agriculture have shown an upward trend under the CAP. Since the 1960s the EU has gone from being a net importer of many temperate agricultural commodities to being the world's second largest exporter. This is a key economic fact in the history of the CAP. Table 2.1 and Fig. 2.2 are both evidence of this economic trend.

Before we consider the effects of this on the operation and fiscal situation of the CAP, it is worth noting that the rapid rise in EU production did not come about independently of the support regime in place at the time. The CAP has been a major factor in the generation and adoption of

Table 2.1. EU self-sufficiency levels (percentage of EU consumption covered by EU production). (*Source*: FAO Production Yearbook, *Agricultural Situation in the Community*.)

		Cereals	Sugar	Butter	Total meat	Beef
1973	EU9	91	91	104	93	99*
1991	EU12	120	128	121	102	107

* 1974.

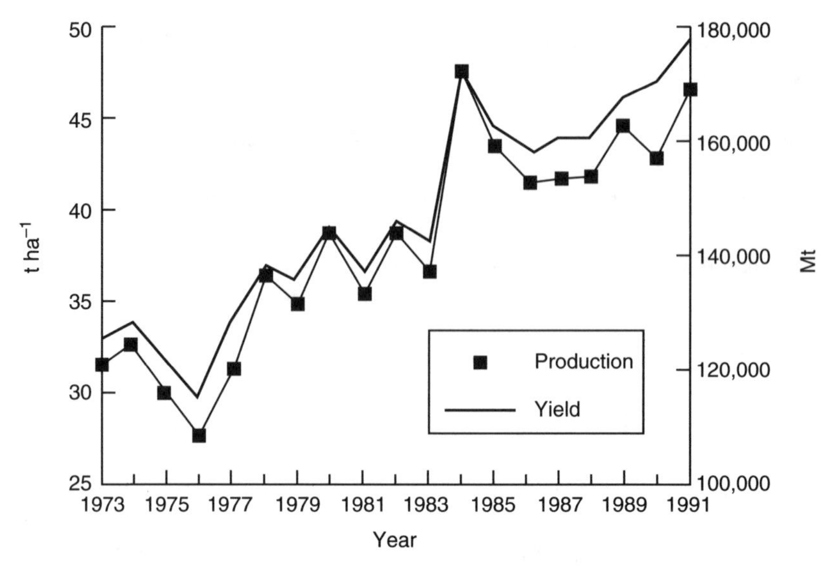

Fig. 2.2. The EU 12 cereals sector: yield and production levels. (*Source: Agricultural Situation in the Community*, FAO Production Yearbooks.)

productivity-boosting innovations and this trend of continually increasing production. The high prices in the CAP have provided an incentive for research and development of these types of innovation. The CoAM could have responded to this trend by cutting institutional support prices in real terms in order to maintain a balance in EU agricultural markets. Fearne (1991) presents the history of decisions by the CoAM to incrementally increase nominal institutional support prices at the annual price review (Table 2.2). The relationship between the Commission and the CoAM in the agreement to ratchet support prices upwards will emerge as one of the main themes of this book.

A generous price support system has two effects on the economics of investment in agricultural innovations. The first is a reduction in the uncertainty of future market conditions; the EU through the intervention price effectively guaranteed to purchase everything that farmers could produce at P_i on Fig. 2.1.

Second, the structure of farming is atomistic. Each farmer is unaware of the relationship between his or her expansion and any downward pressure on price. The CAP, by maintaining a high institutional support price, has resisted this downward pressure on price. Therefore, the price mechanism no longer acts as a disincentive to increase output. Hence production has been encouraged to expand under the CAP.

The economic conditions fostered by the CAP have resulted in a split in European agriculture between an efficient sector which would be able to compete internationally at world commodity price levels, and small-

Table 2.2. Commission price proposals relative to Council decisions (cereals sector). (*Source*: Fearne (1988, p. 108).)

	Council decision		Commission proposal	
Year	ECU	National currency	ECU	National currency
1968	0.0	0.0	0.0	0.0
1969	0.0	0.0	0.0	0.0
1970	0.0	1.5	0.0	1.5
1971	3.5	4.0	0.0	0.5
1972	8.0	9.8	6.5	8.3
1973	5.0	6.5	2.8	4.3
1974	13.5	19.9	11.2	17.6
1975	9.6	13.6	9.2	13.2
1976	7.5	11.4	7.5	11.4
1977	3.9	8.2	3.0	7.3
1978	2.1	8.5	2.0	8.4
1979	1.3	7.4	0.0	6.1
1980	4.8	10.5	2.5	8.2
1981	9.2	10.8	7.8	9.4
1982	10.4	12.1	8.4	10.1
1983	4.2	6.9	4.2	6.9
1984	–0.4	3.2	0.8	4.4
1985	0.0	1.3	–0.3	1.0

Note: The relationship between the ECU (European Currency Unit) and national currencies is discussed on pp. 20–21.

scale farming which would be bankrupted without EU support. CAP subsidies, however, are linked to production. The main beneficiaries of CAP spending have been the most efficient and productive farmers, those who least need support. The often-quoted statistic during the MacSharry reform process was that 80% of CAP spending goes to the most productive 20% of farmers.

The views of agricultural economists on the effect of the CAP on the farm income problem are surveyed in Sections 2.1 and 2.3. However, this is not the factor which has precipitated reform. The growth in EU agricultural production has been the key factor in this respect. The first effect on the operation and direction of the CAP has been persistent surpluses in EU production leading to the growth of publicly purchased and stored stocks. Figure 2.3 provides some evidence of this. These surpluses have been disposed of in two ways, both of which impose a financial burden on the EU budget. Method 1 was that the EU bought the surpluses. Oversupply forced the market price down to the intervention price level. The intervention part of the CAP support system was designed so that the EU could act as buyer of last resort in response to temporary surpluses in markets. However, the

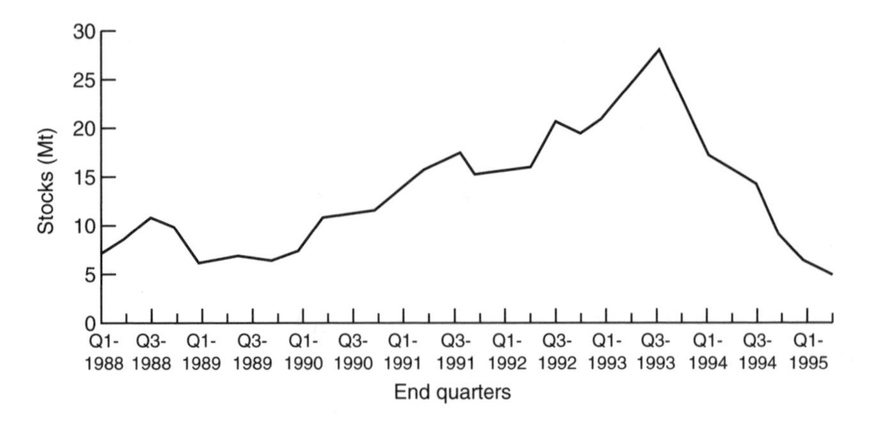

Fig. 2.3. EU cereal stocks, 1988–1995. (*Source*: Ockenden and Franklin (1995).)

consistently increasing production levels in agriculture have often forced the EU to be buyer of first resort. Temporary surpluses could be purchased (through the EU budget) and released at some later point on to the internal market; hence the cost to the budget would be a storage cost. However, as surpluses became entrenched, the main outlet for intervention stocks was the world market. These stocks were sold at tender at the world price; in budget terms this was the same as an export subsidy.

The second method of surplus disposal is through the use of an export subsidy. It has generally been the case that the EU market price has been higher than the world price. In order that farmers receive at least the EU market price, any exports at the world price are subsidized through VESs, which are paid directly from the EU budget. Hence, surpluses disposed of either way impose financial burdens on the EU budget. The shift from net importer to net exporter, which implies a reduction in VIL receipts, plus the cost of disposing of surpluses, had nearly bankrupted the EU by the mid-1980s. In 1984 and 1985, member states had to make an extra one-off payment to the EU budget.

Comparing Figs 2.1 and 2.4 shows that domestic supply has shifted from S_d to S'_d as the EU has moved to being a net exporter of agricultural commodities. If it is assumed that the domestic market price is at P_d, then surplus domestic production is $(Q_2–Q_1)$. Assuming this surplus is immediately exported, then the cost to the EU budget can be represented by the box abcd (i.e. there are no storage costs).

In practice, these budget costs have been exacerbated by exchange rate problems. The absence of a single currency meant that the operation of foreign exchange markets threatened the objective of a common price and a common market for commodities covered by the CAP (see Section 2.2.1, the 'three pillars' of the CAP). Institutional farm support prices (e.g.

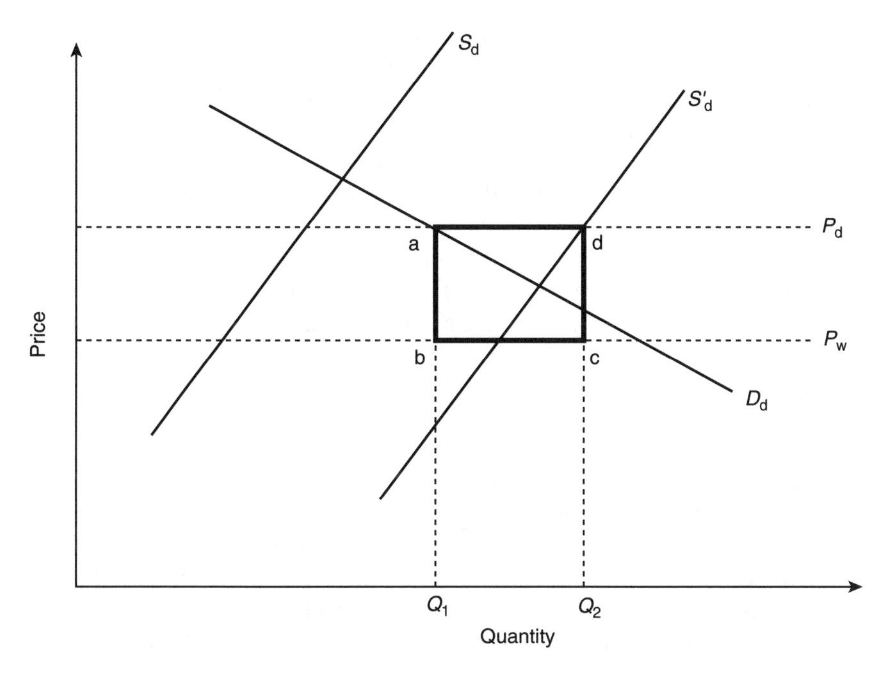

Fig. 2.4. The shift from being a net importer to a net exporter of agricultural commodities.

intervention price) were defined in terms of a common unit, initially in Units of Account (UAs), which were linked to the USA dollar, subsequently the European Currency Unit (ECU), and later the *green* ECU. The problem with defining and setting prices in terms of a common unit is that agricultural markets actually trade using national currencies. As these national currencies change relative to the common unit, so the price of an agricultural commodity in each national currency changes in absolute and relative terms; farmers face unstable prices, which rather defeats the purpose of a price support system.

To counteract this instability, the exchange rates between the common unit of agricultural support prices and national currencies, the green rates, were fixed for some period of time. However, the actual exchange rate in the financial markets continued to vary. The combination of fixed exchange rates for agricultural commodity support prices and floating market exchange rates created arbitrage opportunities and artificially induced trade. Traders could buy agricultural products at a low price (in the weak-currency country) and sell at a high price (in the strong-currency country). To avoid this situation the EU introduced monetary compensatory amounts (MCAs). These were taxes and subsidies applied at the borders to intra-EU agricultural trade. Imports of agricultural commodities to weak-currency countries were subsidized and imports to strong-currency countries were taxed. Periods of exchange rate volatility in the 1970s and 1980s and a reluctance to alter green rates meant that the system of MCAs became extremely complicated and almost destroyed the ambition of creating a common institutional support price in the EU.

The exchange rate mechanism (ERM) of the European monetary system (EMS), established in 1979, eased the tensions on the system of MCAs by reducing exchange rate volatility and the need to change green rates. However, by 1984 the continued strength of the Deutschmark (DM) led to the introduction of the *switchover mechanism*. The adjustment of green rates to take account of a strong DM had led to a reduction in support prices for German farmers and a rise in support prices in weak-currency countries. The switchover mechanism was designed to counterbalance this effect.

The switchover mechanism set the common unit of agricultural support prices, the green ECU, effectively as a proxy for the DM. Realignments in the green rates were made against the DM, and the DM was never significantly revalued against the green ECU (with the associated cut in nominal DM support prices). Instead, all other currencies were devalued (their nominal support prices increased) against the green ECU. This had the same effect as the CoAM agreeing higher support prices, with all the attendant budget problems noted above.

Figure 2.5 shows the cost of the CAP during the 1970s and 1980s, and its share of the total EU budget.

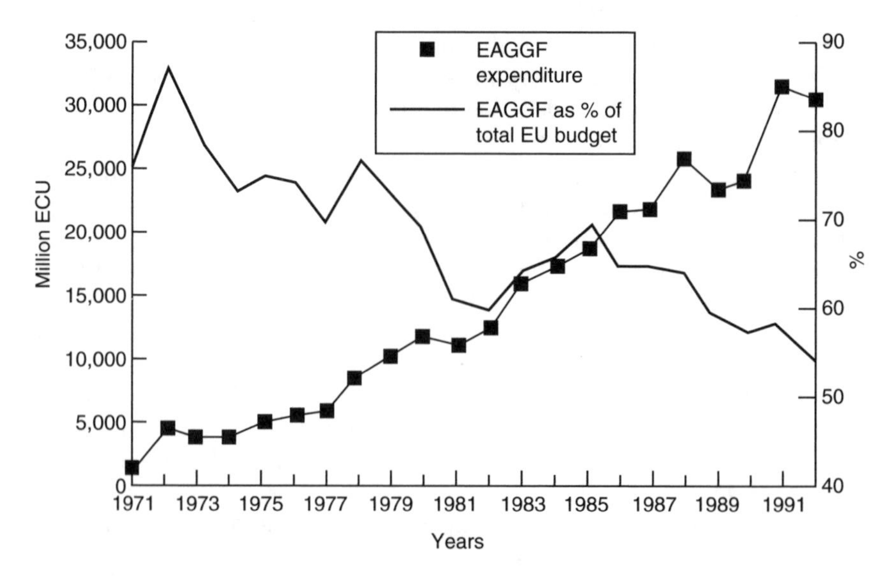

Fig. 2.5. The CAP's domination of the budget. (*Source*: *Agricultural Situation in the Community*, FAO Production Yearbook.)

2.3 A WELFARE ECONOMICS APPROACH TO THE CAP

2.3.1 The Welfare Effects of a Tariff

Figure 2.6 sets out a standard partial equilibrium (PE) analysis of the welfare effects of the imposition of a tariff on a product in a country which imports that product. An *ad valorem* tariff, at rate *t*, would raise the domestic price from P_w to P_t, but lower the export price from P_w to P_t^*. (It is assumed that the country imports enough of the product to affect the world price.) Domestic production increases from S_1 to S_2 in response to the higher domestic price, but domestic consumption falls to D_2.

The costs and benefits of the tariff can be expressed in terms of the areas a–e. Domestic producers gain area a, owing to the higher domestic price, P_t, increasing producer surplus. The higher domestic price makes domestic consumers worse off by area a + b + c + d, this is a loss of consumer surplus. The government receives tariff revenue represented by the area c + e. This is the tariff rate, *t*, multiplied by the volume of imports, where (a) $t = P_t - P_t^*$, and (b) volume of imports = $D_2 - S_2$. The net welfare effect of the imposition of the tariff has three elements:

Consumer loss +	producers' gain +	government gain	
– (a + b + c + d)	+ a	+ (c + e)	= – b – d + e

On the standard assumption (discussed below) that a costless system of compensation is possible, the net welfare effect outcome depends on the balance between the deadweight loss of a tariff (– b – d) and the terms of trade gain (+ e). For a small country there would be no terms of trade gain and hence the net welfare effect of the tariff is negative.

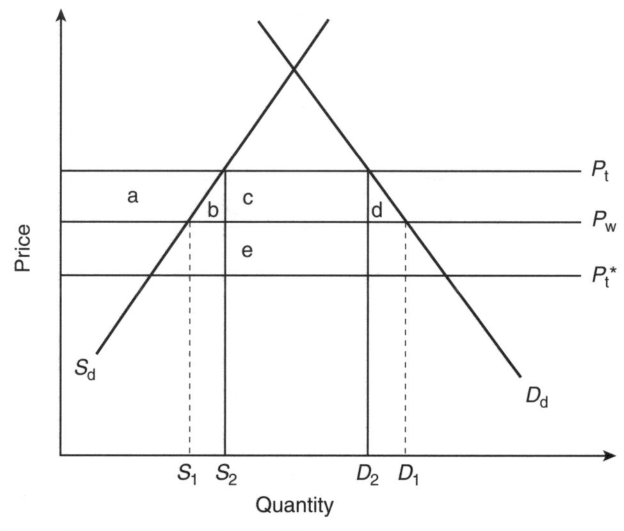

Fig. 2.6. The welfare effects of a tariff.

2.3.2 Welfare Effects of an Export Subsidy

Figure 2.7 shows the welfare effects of the subsidization of the export of a product. The movement in prices as a result of an export subsidy is exactly the opposite of that for the imposition of a tariff (Fig. 2.6). The price in the exporting country rises from P_w to P_s. This is less than the value of the subsidy because it is assumed that the country exports enough of the product to affect the world price, which falls from P_w to P_s^*. The welfare effects are as follows: consumers lose, producers gain and the government loses because it has to spend on the subsidy. The consumer loss is area a + b; the producer gain is the area a + b + c; the government subsidy is area b + c + d + e + f + g. The net welfare loss is, again assuming that a costless system of compensation is possible, the sum of the areas b + d + e + f + g.

Two points emerge from the calculation of a net welfare effect in the case of a tariff and export subsidy. The first is that in both cases there is an associated deadweight loss of welfare. These deadweight losses are the basis of the criticism that domestic price support and export subsidies (the two main instruments of the CAP before the MacSharry reforms) are inefficient devices for transferring income to farmers. Production and consumption are distorted, resulting in a net loss of welfare. In the case of the tariff this deadweight loss is equivalent to the sum of area b and area d. The export subsidy produces a deadweight loss equivalent to areas b and d in Fig. 2.7. This results from a similar distortion of production and consumption. The analysis in Figs 2.6 and 2.7 has provided the base for the prescriptions for CAP reform given by agricultural economists. These prescriptions are outlined in Section 2.3.3.

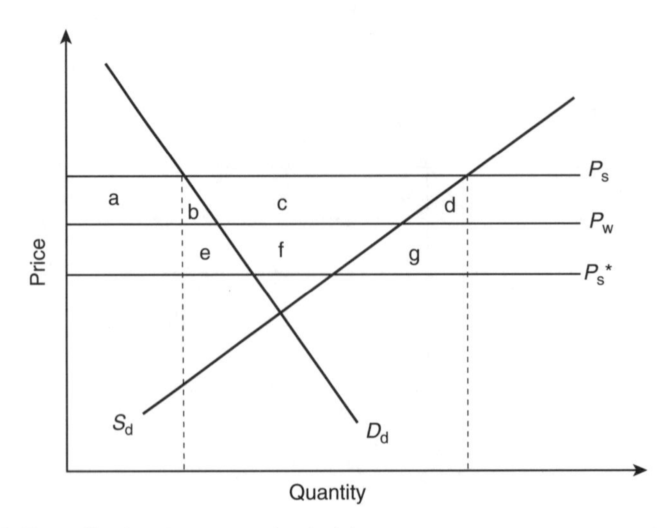

Fig. 2.7. The effects of an export subsidy.

The second point is more general and concerns the assumption that a net effect can be calculated for economic policies which transfer welfare. In the calculations above, it is assumed that the changes in welfare of consumers, producers and the government can be traded off against each other to reach an overall net change in welfare figure. Bonner (1986) imagines a policy which restricts the import of wine and encourages domestic cheese production. There would be more cheese and less wine available for domestic trade and consumption; cheese consumers and producers are likely to be better off, while wine importers and wine consumers are likely to be worse off.

There are two tests which must be simultaneously satisfied in order to claim that the policy has improved social welfare; that is, that there is a positive net welfare effect. The first is the *compensation test.* There are two results for the test, compensation possible and compensation impossible. Compensation is possible if the gainers could compensate the losers (i.e. make them as well off as they were before the change) and still be better off themselves. Second, there is the *bribery test,* which again has two results, bribery possible and bribery impossible. Bribery is possible if before the implementation of the policy, the potential losers can make the potential gainers as well off as they would have been with the introduction of the policy and still be better off themselves in the status quo position.

There are four combinations for the two test results:

		COMPENSATION	
		Possible	Impossible
BRIBERY	Possible	1	2
	Impossible	3	4

Only in situations 2 and 3 is there an unambiguous net welfare result for a policy. The two tests fail to provide unambiguous results in circumstances 1 and 4.[2] This is an important point. It highlights the assumption behind the standard welfare economics analysis of government policies. It is not always possible to arrive at a figure for the net welfare effect of a policy, therefore there is a gap in the analytics of standard welfare economics. Any normative judgement between two economic policies depends on the assumption that the transfer of welfare takes place according to situations 2 or 3. The reliance on this assumption highlights the lack of an account of the government decision-making system, and the role of interests in society in affecting that system, in the standard economic welfare analysis. This provides a rationale for the discussion in the subsequent two chapters of alternative approaches to CAP reforms.

2.3.3 Prescriptions for CAP Reform

Hubbard and Ritson (1991) describe an academic consensus which emerged in agricultural economics' neo-classical, normative approach to the CAP. The following critique is at least as old as the CAP itself. Josling (1973) notes that a similar kind of analysis of USA agricultural support regimes was made in the 1950s and 1960s. The CAP as a price support system is a highly inefficient way of redistributing income to small-scale, low-income farmers. The high support prices benefit larger farms (because, as described, the CAP spending is distributed pro rata to the amount produced). The scheme encourages high-cost, inefficient production. The disposal of the surpluses resulting from increasing EU production has burdened the EU budget and fuelled tension with international trading partners.

Josling (1993) provides empirical estimates of the extent of the deadweight losses associated with different commodity regimes of the CAP during the 1980s. These estimates are expressed as a percentage of the 'pure' welfare transferred (i.e. the welfare transfer that would have been effected without any deadweight losses). Between 1985 and 1990, deadweight losses ranged from 7.5% to 18% of the 'pure' welfare transfer in the wheat regime, from 20% to 30% in the beef regime and consistently over 30% in the sugar regime.

The reform plans which have been proposed by academics throughout the history of the CAP in order to solve the problems identified in this chapter have concentrated on reducing the level of support prices and/or the introduction of some kind of direct income payments scheme. The Wageningen Memorandum of 1973, produced by a group of agricultural economists at a conference at Wageningen University, proposed the shifting of relative prices so that products in short supply should have their prices increased and those in surpluses should have their prices reduced. As most agricultural sectors covered by the CAP had moved into surplus by the mid-1970s, thinking moved on. Josling (1973) proposed a scheme of reducing intervention prices, combined with direct income payments linked to output. The level of the direct income payments would be insufficient to provide inefficient farmers with a reasonable standard of living. These inefficient farmers would get discretionary income supplements limited either in duration or to the incumbent generation of farmers. Thus the inefficient producers would not be penalized but their replacement by future generations would be discouraged. Koester and Tangermann (1977) propose a direct income payments system which would be *decoupled* from production. They argued that this would be a more efficient way of redistributing CAP support spending in favour of small-scale, low-income, marginal farmers.

Marsh (1977) proposed a system of common trading prices set near to the world price. Member states would be able to offer their national farmers whatever price they wished, but when CAP commodities were traded between member states the price charged would be the common trading price. If one member state wished to maintain an internal price above the common trading price then any exports from that member state would have to be subsidized (to the difference between the internal price and the common trading price) by that member state. This proposal took more account of member states' desire to support national agricultural support prices than did the reform proposals noted above. The proposal informed much of the later debate on the renationalization of the CAP (Wilkinson, 1994). Further, unlike the Josling (1973) proposal, it anticipated that it would be the budget issue rather than the inefficiency of the high price supports as a solution to the farm income problem which would force the reform of the CAP.

2.4 CONCLUSION

The farm income problem stems from the fixity of resources in the agricultural sector. There is an oversupply of resources in farming, especially labour, and this keeps returns on resources employed there chronically low. The CAP is an inefficient and expensive way of trying to rectify the farm income problem because it is a price support system: inefficient because there is a deadweight loss in economic welfare due to the distortion of consumption and production; expensive because, as Section 2.2 shows, CAP spending is linked on a pro rata basis to production. Under the CAP, the EU has gone from being a net importer to a net exporter of most agricultural commodities. This trend has contributed to the CAP's domination of the EU budget. The prescriptions of agricultural economists for CAP reform have generally been aimed at correcting the inefficiency of the policy, although some have recognized that the budget expense of the CAP may be a stronger motivating factor for policy-makers seeking reform.

NOTES

1. Tracy (1989) provides the most complete histories of these pre-CAP national support regimes.
2. Circumstances 1 and 4 arise because utility possibility curves can intersect. See Bonner (1986) for a full explanation.

Chapter 3

The History of CAP Reform

The chapter starts by arguing, in Section 3.1, that the neo-classical, normative welfare economics approach to the CAP, outlined in Chapter 2, ignores the policy process as a factor in the reform of the policy. This book holds that an understanding of CAP reforms must be based on an explicit account of the policy process which leads to the enactment of a particular reform.

The histories of the following reforms of the CAP are described in Sections 3.2–3.4: the Mansholt Plan, the milk quota reforms and the introduction of the stabilizer regime. The descriptions include the policy processes involved as well as an economic analysis of the reforms. These histories show that with one exception, the input of agricultural economists has had a negligible effect on the reform of the CAP.

Section 3.5 draws parallels in the episodes of CAP reform before the MacSharry reforms of 1992. These parallels between the three reforms described in this chapter are used in Chapter 5 to support a common analytical framework for CAP reforms, a framework that is then applied to the MacSharry reforms in Chapters 6 and 8.

3.1 POSITIVE VS. NORMATIVE APPROACHES TO THE ANALYSIS OF CAP REFORM[1]

The Mansholt Plan of 1968 seems to have been informed to an extent by the analysis of factor immobility provided by agricultural economists (see Chapter 2). The proposals for structural adjustment were a response to the conclusion that the lack of labour mobility (out of agriculture) was a central factor in the farm income problem. Further, the farm income problem was accepted as the main rationale for government intervention in agriculture. However, the description of the history of CAP reforms provided in Sections 3.2–3.4 points to a limited influence for the analysis of agricultural economics in the reform process. The prescriptions of agricultural economics (described in Section 2.3) seemed to have had little influence on, or even relevance to, the agenda of CAP policy-makers in the 1980s. As Pelkmans (1985) termed it, there was a 'dialogue of the deaf' between policy-makers and agricultural economists on the subject of CAP reform.

The neo-classical, normative agricultural economics approach to the analysis of the CAP can explain the emergence of pressures for reform. Specifically, budget crises in 1983/84 and 1987/88 are identified in Sections 3.3 and 3.4 as important in producing reforms of the CAP. However, the aim of this work is to provide an understanding of the MacSharry reforms of the CAP. Neo-classical economics can nominate pressures which have built up owing to the operation of the CAP as causes of reform, and it can also provide an economic analysis of the effects of a CAP reform. What the neo-classical approach has been unable to answer is why the dictates of economic rationality have had such a limited effect on the CAP reform process; why there has been a dialogue of the deaf.

A full understanding of the MacSharry reforms requires some answer to each of the questions posed at the beginning of the book: Why were those reforms enacted? When were they enacted and why were they enacted in the way that they were? To provide such answers requires an analysis of the policy process: why certain pressures are important, and in certain places and at certain times in the decision-making system. For example, the neo-classical, normative agricultural economics approach of Chapter 2 has no real way of answering why the policy process responded with a CAP reform to the budget problem rather than the farm income problem. It is the claim of this chapter that an understanding of how the CAP decision-making system responds to pressures and why reforms are enacted, when they are, and in the way that they are, requires a positive analysis of the CAP decision-making system.

There is the danger of material determinism through the *ex post* argument that the reforms that were enacted were inevitable because of the dire financial situation of the EU budget. This can lead to the view that the CAP decision-making system is a black box which responds to objective, material pressures with inevitable reforms. Material forces cannot, of course, be avoided. They form the context of any CAP reform. However, policy-makers do not view them as an agricultural economist would. These material forces do not motivate an ambition for economic efficiency among policy-makers, and the interest of economic rationality and efficiency is not obviously represented in the CAP reform process.

Marsh (1985, p. 120), when discussing future potential for CAP reform, stated that 'it is not necessary to be a cynic to believe that ministers, even prime ministers, may prefer the confused unsatisfactory packages of stop-gap measures to shore up the CAP rather than *reform* in any fundamental sense'. A positive approach to the study of CAP reform is aimed at answering the question of why the CAP decision-making system responds in such a way to pressures for CAP reform. This is a separate question to why budget pressures emerged through the history of the CAP.

In the 1980s, there was an intellectual movement towards positive approaches to CAP decision-making motivated by the ambition to explain when, why and how the CAP was reformed. The argument was that neo-classical, normative agricultural economics failed to analyse CAP decisions as the product of a system. The decision-making system of the CAP did not respond in an inevitable way to the economic and financial pressures or to the logic of economic rationality. Hagedorn (1983) wrote a seminal paper in encouraging a positive approach to studying CAP reforms and a more complete view of the CAP decision-making system.

Hagedorn (1983) sets up the view that agricultural policy exists in two domains. His claim is that a strict agricultural economics approach analyses only one domain. As such it cannot either fully explain the reform process or provide policy advice of practical relevance. The two domains of agricultural policy are market co-ordination and political co-ordination. The strict agricultural economics approach, outlined in this chapter, seeks to understand the history of the CAP in terms of the situation in agricultural markets and its consequences for the EU budget. This is the market co-ordination domain. This approach ignores the political co-ordination domain of agricultural policy, 'the various institutions regulating the politician's decisions and the policy process, for example: elections, collective action, constitutional agreement, interest groups, bureaucracy, negotiations, coalitions, etc.' (Hagedorn, 1983, p. 304). Hagedorn's paper makes three main points: (i) the operation of the CAP directly affects the agricultural markets of the EU and the global commodity markets; (ii) decisions about the operation of the CAP are the product of a political system; (iii) agricultural policy decisions, including decisions to change it, are both affected by, and affect, that political system. CAP decisions politically co-ordinate those parts of the state and polity which have an interest in agricultural policy. Hence, CAP reforms cannot be fully understood without some appreciation of this political co-ordination domain of agricultural policy.

Hagedorn (1983) was followed by Pelkmans (1985), Senior-Nello (1984) and Hagedorn (1985). All attempt to understand the two domains of agricultural policy-making noted in Hagedorn (1983). Accounts of the CAP reforms of 1984 and 1988 have been written by agricultural economists, but using approaches which try to take account of the decision-making process which produces CAP reform. Tracy (1989), Moyer and Josling (1990), and Hubbard and Ritson (1991) account for stabilizer reforms using approaches beyond strict normative agricultural economics. (A description of the stabilizer regime is given in Section 4.2.) Petit *et al.* (1987) similarly explain the introduction of milk quotas in 1984.

The intellectual development from a normative agricultural economics approach has been part of a broader sweep in economics generally, in which economists have attempted to give a positive and more substantial account of government and public policy decision-making systems. The models developed have been at a very abstract level, but agricultural support policies are mentioned as empirical examples. Downs (1957) and Stigler (1960) were pioneers in this field. The public choice paradigm described in Chapter 5 is part of the answer to this demand for a positive theory of how governments and the wider political system work.

3.2 REFORMS OF STRUCTURAL POLICY

The policy process associated with the Mansholt Plan of 1968, also known as 'Agriculture in 1980', for a structural adjustment in European agriculture is included in this chapter for four reasons. First, it was the first CAP reform proposed. Second, the lack of action taken by the CoAM at the time became typical of what happened to future reform proposals. Third, the Mansholt Plan foreshadowed many of the problems which were factors in the reforms of the CAP in the 1980s. The plan is the one significant example of the analysis of agricultural economists, prior to the MacSharry reforms, having some input into the reform process.

The fourth reason for interest in the Mansholt Plan is that the failure to adopt fully the plan's structural adjustment programme is an example of how the question *for what purpose does the CAP exist?* has never been properly answered. After 1969, and the failure to establish a structural policy, price policy alone has been used to try to satisfy two answers to this question. There is the *technocratic* answer and there is the *social policy* answer. The technocratic premise for the existence of a CAP is that agriculture in Europe should be made efficient and competitive in terms of world agriculture. The alternative premise for the CAP is that agriculture is a declining industry and agricultural policy is about managing that decline in a socially acceptable manner, so the CAP should be a social policy. Chapter 2 illustrates how price policy and its consequences have dominated the history of the operation of the CAP.

3.2.1 Background

This section describes and evaluates the debate over the structure of agriculture as it existed in the EU between 1958 and the implementation of the directives known collectively as *mini-Mansholt* in 1972. This section also describes what structural policy is, and how it contrasts with price policy as a method of intervention in agriculture. It also details the six structural policies which were operating at the national levels before the introduction of the CAP. Section 3.2.2 outlines the Mansholt Plan of 1968 (which is different from the 1960 plan also known as Mansholt), what it proposed and what was eventually enacted, concentrating its focus on the difference. Section 3.2.3 briefly outlines what happened to the structure of agriculture in the European Community up to 1980, the period which the Mansholt Plan covered. This provides a background to the kind of pressures for reform which existed in the 1980s and which led to the enactment of CAP reforms in 1984, 1988 and 1992.

Structural policy is something which has existed in the member states of the EU from the Second World War onwards. The term is generally used to describe attempts to make the agricultural sector more efficient in its use of land, labour and capital inputs. It concentrates on the supply side of agricultural markets; in particular, the capacity of the agricultural industry to supply efficiently. The demand side of that equation – that is, what quantity is bought and at what price – comes under the purview of price policy. The national policy-makers' focus on structural policy was induced by the exigencies of wartime and the immediate period thereafter. The capacity of European agriculture to supply the demand for food was a genuine concern in post-war Europe. According to Hallett (1968), structural policy consisted of the consolidation and amalgamation of the fragmented holdings prevalent across continental Europe. Tracy (1989) provides an account of the history of this fragmentation. Consolidation refers to the bringing together of scattered strips within the same farm holding, whereas amalgamation is the grouping of smaller, separate farm holdings into a larger single enterprise. However, the term 'structural policy' has developed a wider meaning to encompass labour reforms, such as the retraining of farmers when leaving agriculture, or pension inducements for older farmers, and capital usage measures, e.g. tractor grants.

At the time of the Mansholt Plan, France had the most developed structural policy of the six founder members of the EU. A 1945 government report estimated that 10 million hectares (Mha) required *remembrement* or consolidation, of which just under half was completed by 1965 (Neville-Rolfe, 1984). However, consolidation does not itself affect the number or average size of holdings. The 1960 Loi d'Orientation Agricole established SAFERs (sociétés d'aménagement foncier et d'établissement rural). SAFERs were designed to extend structural policy to create new holdings as well as reshape existing ones. They bought agricultural land offered on the market and resold it to 'suitable' purchasers within 5 years. They enjoyed pre-emption rights in the purchase of any land freely offered but had no compulsory purchase right. These powers were strong, contrasting with the relatively weak powers enjoyed by the Agricultural Land Commission in the UK. The declared aim of SAFERs was to increase the number of viable family holdings, as opposed to creating an internationally efficient scale of farming in France. This reflected public opinion, which was against the extension of very large holdings. As Butterwick and Neville-Rolfe (1965, p. 553) noted, 'The purposes of SAFERs are at least as much social as economic'.

The need for structural reform in UK agriculture was first explicitly recognized in the 1967 Agriculture Act, which was founded on statistics that showed that half of the holdings in England and Wales could support only part-time farming and a further quarter which were run full time were too small to be viable. The main provisions of the Act were small retirement pensions for older farmers and amalgamation grants. Hine (1973) demonstrates that the scale of these provisions was too small for the Act to have any significant effect.

Hallett (1968) observed that the size of the viable farm is constantly being revised upwards. This insight may provide an explanation of why the farm problem has never died out: young people entered farming in the 1960s on a scale which was then viable but which has subsequently become non-viable. The target of the modern, efficient farm was, and is, a moving one.

Germany, Denmark and the Netherlands had similar schemes by the mid-1960s; structural policy occupied the agenda of national governments. This next section shows that the Mansholt Plan failed to produce a consensus that the issue of the structure of European agriculture should be dealt with at a European level. Further, this failure is characteristic of the CAP decision-making system.

3.2.2 The Mansholt Plan of 1968

The 1960 Mansholt Plan provided the basic principles of the CAP. It was based on Commission discussions after the Stresa conference and included reference to structural policy as well as price support. Neville-Rolfe (1984) argues that political expediency meant that this section of the Commission's plan was dropped until the completion of the common market in agricultural products; that is, the whole project of a CAP would have been lost if common market and structural proposals had been made simultaneously. This view is supported by a document reproduced by Weigall and Stirk (1992, p.136) entitled 'Sicco Mansholt on the reform of the CAP', written in 1970. Mansholt argues there that the Mansholt Plan (1968) was not proposed earlier by the Commission (i.e. in parallel with the completion of the common market for agricultural produce) because the member governments 'wouldn't hear of it'. The completion of the price support regime of the CAP was consuming all available political energies. Thus, as soon as the single market was completed for most goods in 1968, the Commission introduced the Mansholt Plan, with the view that the CAP would be ineffective so long as the majority of holdings were incapable of providing an adequate income (even with price support). It was politically easier to gain common agreement for price policy than for structural policy, because price policy promised to raise the support price for the majority of farmers in the EC, whereas the Mansholt Plan aimed to remove 5 million farmers from agriculture.

There were four main aspects to the Mansholt Plan. The first was what Neville-Rolfe (1984) describes as a 'forthright' statement against price policy described by Mansholt (Weigall and Stirk, 1992) as 'based on consensus politics rather than economics'. 'Agriculture in 1980' predicted (in 1968) that such a price policy would lead to a structural surplus which would soon bankrupt the Community. There was no alternative to structural reform to create 'modern, viable farms', Mansholt argued; 'financing farms with five cows is tantamount to financing chronic destitution'.

The second main aspect of 'Agriculture in 1980' was the plan, mentioned above, to move 5 million agricultural workers out of farming by 1980. Three million could be removed by pensioning off farmers over the age of 55, and the remainder were to be taken by the creation of 80,000 new jobs per annum in the less industrialized areas of the EU.

Aspect three of the 1968 Mansholt Plan was the 'creation of agricultural enterprises of adequate economic dimension' (Neville-Rolfe, 1984, p. 302). Grants were to be made available to production units (PUs), where PUs were defined as production on a scale laid down for each product group. This would encourage pooling or sharing of farm resources on a product-by-product basis – it was known at the time as partial amalgamation. Also in 'Agriculture in 1980' was the concept of the modern agricultural enterprise (MAE), to encourage the complete amalgamation of farms by making grants available to units covering the production of several commodities.

The fourth aspect was to extend the scope of common financing of structural measures. The EAGGF would fund half of the total expenditure of national programmes to take people off the land, compared with the situation of funding not more than 25% of a few selected programmes.

The reaction in the CoAM to the Mansholt Plan was negative; each member state found something unacceptable in it. The French government worried about the control the Commission would be gaining over farm structures; it argued that the power to intervene directly in agriculture, as opposed to indirectly through the price mechanism (though the end result may be the same), was a much stronger one. The MAE and PU sections took up half of the document, encouraging the view of one Gaullist deputy of a 'frigid technocracy' (Neville-Rolfe, 1984, p. 304) emerging in the Commission. The West German government's reaction was equally hostile. Two arguments were made. The first was that to define the efficient scale of inputs, commodity by commodity in a mixed farm, does not ensure the efficient mixture of those inputs. Most factor inputs will have a certain transferability across sectors, in which case the scale for each individual commodity did not have to be as high as the Commission proposed. The second and more fundamental point was an unwillingness to bear the cost of the alteration of the farm structures of other countries. The Mansholt Plan contained certain assumptions about economic growth and the migration of workers leaving agriculture up to 1980. The German government was worried about the potential cost of structural policy should those assumptions not be correct. Joseph Ertl, the Minister of Agriculture, proposed to maintain prices but limit the level of surpluses by using quotas as discussed in Sections 3.3 and 3.4; the demand for supply-side management through quotas has been a constant theme in the German government's attitude to the

CAP over 20 years. Meanwhile, existing national policies would continue to smooth the adjustment taking place in agriculture.

Mansholt addressed the Council of Agricultural Ministers in October 1967, trying to instil a sense of urgency to his view that the aim of parity with non-farm incomes pursued solely by price policy would eventually bankrupt the Community through enormous surpluses. However, such dire warnings failed to pave the way for the agreement of a structural policy along the lines proposed in the Mansholt Plan. The political struggle of completing a common price policy in agriculture had just finished and the member states were far from receptive to a proposal calling for fundamental reform of that policy. The Vedel Report (1969) was commissioned in 1967 by Faure, the French Minister of Agriculture. The report addressed the problem of creating an agricultural sector capable of adapting production to the needs of the market at competitive prices (Marsh and Ritson, 1971). Vedel shared Mansholt's technocratic view of agricultural policy. The Vedel Report proposed price support cuts to bring agricultural markets near to balance, as well as a form of set-aside to take 11 or 12 Mha of land out of production (which bears some similarities with the MacSharry reforms). The report was similar in philosophy and tone to Mansholt. The French government delayed its publication by 3 months to placate criticism and immediately distanced itself from the Report's proposals or conclusions. The reaction to its substance, in the country with the most developed structural policy in the Community, gives an example of the climate of opinion into which the Mansholt Plan had been pitched.

The failure of the Mansholt Plan to gain any sort of political momentum in the CoAM led the Commission to pare down the proposals; the revised plans were collectively known as mini-Mansholt and were published in November 1969. The background to this version of the plan were the negotiations for the final arrangements for the financing of the CAP, the difficulties of agreeing a price package for 1970/71 and the problems of nascent surpluses. All of these were to be considered at the December 1969 summit, held in The Hague. The mini-Mansholt Plan proposed reducing support prices for 1970 in sectors in structural surplus, i.e. grain, milk and sugar. The funds saved would be used to finance limited Mansholt restructuring. The plan also insisted on a structural solution to persistent surpluses before there was any final agreement on the financing of the CAP.

The Hague summit concluded agreement on the common financing of the CAP and the 1970/71 price package. This was done without serious discussion of structural reform. Summit packages normally grow in size and complexity. Extra issues can be traded off against outstanding issues in order to achieve an overall compromise. In this case, the introduction of structural policy to the agenda was not required to achieve agreement.

3.2.3 The Effects of the Failure to Agree the Mansholt Plan (1968)

Some structural directives (Reg. 159/72, Reg. 160/72, Reg. 161/72) were enacted in 1972 (Rolfe (1984) correctly calls them 'very modest'). These were agreed when the Commission agreed higher support prices for 1971/72 than it had originally proposed. This was the beginning of the history of the CAP, the need for agreement in the annual price review dominating the agenda of the CoAM. Any different ambition for the CAP had to be subordinated or somehow linked to this annual price review.

One consequence of the failure to agree a common structural policy at the EU level was that the national structural and social policies came to be used by member states to 'compensate domestic agriculture in the case of 'insufficient' price decisions at the Community level' (Schmitt, 1986, p. 340). This is a classic moral-hazard problem. The EU has committed itself to bearing the cost of an outmoded farm structure through the price supports of the CAP. There exists no incentive for any national member state to incur the financial and political cost of farm structure reform. Instead, the financial burden of an increase in agricultural production in a member state is externalized through the EU budget. Therefore, so long as a member state is a net beneficiary of the EU budget, there exists an incentive to use national structural policies to boost national production; the costs are spread through all member states, but the benefits are concentrated in that particular member state.

Neville-Rolfe (1984) argues that the effect of these structural directives has been fairly marginal; the Commission was still producing draft regulations recognizing the failure of structural policy in the EU, especially in more backward areas, in 1983. The three directives of 1972 mark the high point of the EU's attempt to reform the structure of European agriculture. Reform since then has concentrated on the problems of the CAP price policy.

Figures given by Clout (1984) suggest the failure of the various national schemes to address the question of the structure of farming. From 1970 to 1980, the number of farms of 1 ha or more in the EU dropped by 14%. Between 1975 and 1980 the average size of farms rose from 15 ha to 18 ha. Only those in Italy and Greece were significantly smaller than this average size, and only those in the UK and the Paris basin are significantly greater. Clout (1984) cites France as an example of how national structural policies have failed to produce a 'revolutionary change in farm size'. The ownership of one-third of all French farmland from the 1960s to 1980 was transferred through SAFERs. However, the implications for increasing farm size by encouraging amalgamation were limited because of complicated regulations to safeguard family farming. This again brings attention to the question of agricultural policy at national or EU level: does it serve the purpose of technocratic efficiency or that of social ends?

By the 1980s, the issue of structural reform had disappeared from the agenda of agricultural policy-makers in the EU, primarily because of the persistently high unemployment levels in the non-agricultural sectors of the EU economy. There was little point in moving uneconomic farmers into urban unemployment.

3.3 THE MILK QUOTA REFORMS 1984

The development of the dairy sector in the EU is closely linked to the debate over structure. The milk market was the first area to display the characteristics of structural surpluses predicted in the Mansholt Plan. By the end of the 1960s, the Community was producing more milk 'than could be consumed internally at prevailing prices' (Marsh and Ritson, 1971). The German Agriculture Minister's proposal of quotas was aimed specifically at the milk sector. The price review for 1969/70 was upset by the emergence of milk surpluses. The CoAM did not want to cut prices in response – a position which became familiar through most of the 1970s and 1980s. The issue was eventually resolved by agreement on the premium provided for the slaughter of dairy cows, which Marsh and Ritson (1971) describe as having a 'negligible' effect on total milk production, at best slowing the rate of increase. The key point here is that the problem of milk surpluses and the idea of milk quotas as a solution had existed for 15 years before they were actually introduced in 1984. The relevant questions are why the solution was so long in coming, and what the factors were that determined that milk quotas became the 'unavoidable' solution in 1983/84 but not in 1969. It is not just hindsight which predicts the burgeoning milk surpluses through the 1970s and early 1980s; surpluses were emerging in 1969 and were predicted to grow substantially. The milk sector shows how the agricultural policy-making apparatus of the EU produces solutions or CAP 'reforms' only when the problem is immediate; that is, there was a possibility that the Community could have run out of money in 1984 in the absence of measures to curb the cost of financing milk surpluses.

3.3.1 Background to the Milk Quota Reforms

The rising budget costs of the milk sector had resulted in the introduction of a co-responsibility levy (CRL) in 1977, which was initially set at 1.5% of the value of production, later raised to 2.5%. The early 1980s saw the budget of the milk sector grow rapidly; in 1982 it was ECU 3.3bn, in 1983 ECU 4.4bn, and by late 1983 it was being predicted to rise to ECU 5.8bn in 1984. Petit *et al.* (1987) have termed this the 'cost of not taking a decision'. Their thesis is that the decisive factor in removing the status quo as an option for the CoAM was the rising budget costs of the milk sector. The European Council meeting at Stuttgart instructed the Commission to provide some plans for 'concrete steps' to control agricultural expenditure by August 1983.

3.3.2 Reform Process

In July 1983 the Commission submitted COM (83) 500 to the CoAM. COM (83) 500 (Commission of the European Communities, 1983b) included proposals for a system of farm-level quotas for milk production and a special levy on milk produced by intensive methods. This plan was controversial within the Commission, passing by only one vote in the College. Petit *et al.* (1987) state that several commissioners preferred price cuts to quotas and others thought COM (83) 500 was overly concentrated on milk, pushing 'Mediterranean' products in the CAP down the agenda. The CoAM set the Commission a timetable to turn the COM (83) 500 plan into a dossier for consideration at the European Council meeting in Athens on 4–6 December 1983. By this time, quotas had gained some kind of grudging acceptance because of the Commission's prediction that the alternative scenario was a 12% price cut to stabilize or reduce production.

Even though 2 half-days out of a total 48-h summit were devoted to the dairy issue, there was a failure to achieve unanimous agreement on COM (83) 500. The major obstacle was the issue of quotas. This failure precipitated a sense of crisis in the EU because of the genuine threat of the budget simply running out of money.

Petit *et al.* (1987) comment that the French Agriculture Minister, Michel Rocard, who was President of the CoAM in the first 6 months of 1984, proved diligent and successful in establishing agreement in the CoAM. This was similar to the effect of the West German presidency in the first half of 1988 (Section 3.4.4).

In March 1984, the CoAM agreed a package including milk quotas, the dismantling of MCAs (another part of COM (83) 500), and the agricultural prices for 1984/85. This agreement was made conditional on the European Council sorting out certain agricultural and budgetary issues. The European Council of 19–20 March failed to do this; it postponed the issue of the UK budget rebate (eventually agreed at the Fontainebleau summit in July) and could not decide whether quotas should apply in Ireland when the Irish government claimed a vital national interest to obtain special dispensation.

The Commission brought new proposals to the CoAM which formalized the conditional agreements of 5–6 March and added measures which dealt with Irish milk quotas. The CoAM finally reached agreement on the introduction of milk quotas with special treatment for Italy, Greece and Ireland. The overall package has been described by Petit *et al.* (1987) as 'strikingly close' to COM (83) 500.

Domestic politics also intervened in the relationship between the CoAM and the European Council on the issue of CAP reform. Petit *et al.* (1987) describe a strong rivalry between President Mitterrand of France, and Rocard, which manifested itself in a rivalry between the European Council (of which Mitterrand was president for the first 6 months of 1984) and the CoAM (of which Rocard was president in the first half of 1984).

The competition was muted by their being members of the same government. Their relative offices in the French government meant that in the early period Rocard could not risk outright support for milk quotas for fear of being overruled by the President. He considered CAP reform to be an issue of such political importance that it had to be considered by the European Council before the CoAM could reach agreement. It was only when the Athens Summit failed to reach a final solution to the surplus problem that Rocard brokered an agreement in the CoAM around milk quotas. Against this background the conditional constraint imposed on the CoAM decision of 5–6 March could be interpreted (as suggested by Petit *et al.*, 1987) as a gamble by Rocard to claim the credit for the solution to the dairy issue. Rocard gambled that the European Council would be unable to reach agreement on the 5–6 March consensus in the CoAM. The CoAM provided a final agreement a few days later.

Even if this interpretation is correct, the president of the CoAM still felt that on a major CAP reform the European Council had to discuss and reach some kind of broad consensus before the CoAM could reach a final agreement. It is the function of the CoAM to make decisions about the CAP and this power was effectively abnegated by making the 5–6 March agreement conditional on the European Council. Agreement, formal or informal, in the European Council was made a necessary condition for CAP reform.

3.3.3 The Economics of Milk Quotas

The imposition of a quota system restricts domestic supply. This could have the effect of increasing aggregate farm revenue. In most cases the demand for agricultural products is inelastic. Further, the EU market can be considered to be isolated from the world market. Any increase in domestic supply will tend to bring forth a more than proportionate fall in price, reducing aggregate farm income. A policy of restricting supply could, in such circumstances, increase aggregate farm income.

There are two basic disadvantages associated with a quota scheme. First, they introduce rigidity into the farming sector, preventing expansion by low-cost producers and preserving high-cost ones. Initially milk quotas in the EU were not transferable unless the farm was sold, leased or transferred by inheritance. However, after 1987 the sale of milk quotas was officially sanctioned. This had the effect of reducing the rigidities imposed by quotas: the more efficient producers could buy the right to produce from the less efficient. However, while small farmers who could sell their quota would gain in income terms, the purchase of the right to produce is a cost which represents an impediment to larger farm size as a method of improving the incomes of some producers. There is to some degree a trade-off between improving the productive efficiency of a quota system and improving its efficiency as a method of income support.

The second basic disadvantage of a quota system is the administrative cost of quotas, which becomes greater both as the number of producers involved increases and as the scheme attempts to introduce flexibility. This factor is often ignored by economists but has been a large issue in the operation of the EU's milk quota system. At the time of the MacSharry reforms, Italy had still not implemented the quota system, citing administrative problems as the reason.

3.4 THE ESTABLISHMENT OF THE BUDGET STABILIZER REGIME

This section is divided into five. The first subsection looks at the background to the reform process which resulted in the enactment of the stabilizer regime. The next three subsections cover the three distinct phases of the policy process which led to the stabilizer regime. The fifth looks at the economic analysis of these reforms.

3.4.1 Stabilizer Reform Process: Background

The stabilizer regime was enacted on 13 February 1988 by the European Council. It was part of the overall agreement of a 5-year budget plan for the EU. Subsequently, this package of budget measures came to be known as Delors I. Delors II was enacted at Edinburgh in December 1992 to provide another 5-year budget guideline (see Section 3.4.2) for the EU. These packages were so called because of the influence the Commission President, Jacques Delors, and his cabinet had in their construction and enactment. Delors was also instrumental in the MacSharry reforms (see Chapter 6).

The budget agreement in 1988 included a new fourth resource for the EU: a strict limit on growth of EAGGF expenditure to 74% of the growth of EU budget spending and a doubling of funds available for regional policy. Chapter 4 contains a description of how the stabilizer mechanism worked in practice and became a factor in the MacSharry reform process.

The reform process took a year. It started with the adoption of COM (87) 100, *Making a Success of the Single European Act: a New Frontier for Europe* (Commission of the European Communities, 1987a) by the College of the Commission. This 'reflections' document noted the need for increased budgetary resources and increased budgetary discipline to help the financial situation of the Community. Agriculture was (and still is) the largest item of expenditure in the EU budget and was therefore central to any medium-term budget agreement.

COM (87) 100 described an emerging budget crisis. The 1984 Fontainebleau budget agreement had not controlled the growth of EAGGF spending. Instead of the 2% per annum prescribed growth in CAP spending, between 1984 and 1987 the average per annum increase had been around 18% (*AgraEurope*, 26 June 1987). Only clever accounting had concealed a budget shortfall of ECU 4–5bn in 1987 (Moyer and Josling, 1990).

COM (87) 100 contained a serious estimation that the EU would run out of money to meet its obligations sometime in 1988. The cereals sector of the CAP had been displaying the greatest growth in expenditures. Strong EU production levels combined with a weak dollar after the Plaza Accord of 1985 had increasingly burdened the EU budget. Cereal support budget costs in 1987 were forecast to reach ECU 5.9bn in 1988, close to the ECU 6.5bn spent in the dairy sector. The dairy sector had triggered the milk quota reforms; in 1984 the dairy sector spending was over ECU 5bn, with cereals expenditure under ECU 2bn.

Three distinct phases in the reform process can be distinguished. Phase 1 was the development of the reform proposals and their submission to the CoAM on 22–23 September 1987. Phase 2 was the subsequent negotiations in the CoAM culminating in the collapse of the Copenhagen Summit in December 1987. The third phase culminated in the agreement of the stabilizer regime on 13 February 1988.

3.4.2 Reform process: Phase 1

COM (87) 100 was a reflections paper, followed by COM (87) 101 (Commission of the European Communities, 1987b) in February 1987, which focused on the need to control expenditure. Automatic budget stabilizers were proposed to try to limit the growth in agricultural support spending. These budgetary stabilizers were to work as follows. A spending guideline would be set for each commodity. If this was exceeded during the year then the Commission would have powers to adjust prices, intervention levels, etc. to keep spending within that guideline. A budgetary overspend in any one financial year would prompt action by the Commission in that year to keep total spending within the guideline. The general theme of the document was the need to correct the deficiencies in the EU's agricultural policy decision-making system which allowed the CoAM consistently to spend beyond the previously agreed guideline.

The development of the reform proposals in the 6 months after COM 100 and COM 101 is characterized by Moyer and Josling (1990) as being driven by initiatives from Commissioners and their cabinets rather than the Directorates-General. This was the zenith of Delors's influence within the Commission (Ross, 1994, 1995; Grant, 1995b). Henning Christophersen (Budget Commissioner) and Frans Andriessen (Agriculture Commissioner) saw the problem of the agricultural budget as part of a wider budget issue of the future financing of the EU and ultimately the future priorities of the EU. This Delors–Christophersen–Andriessen axis was the driving force behind the reforms; Moyer and Josling (1990) describe an inner circle dominated by these three Commissioners and their cabinets. The inner circle was an *ad hoc* grouping composed of these Commissioners, their advisers and key officials. A similar type of grouping was involved with the MacSharry reforms (see Chapter 6).

The milk quota proposals were much more closely associated with the senior levels of DG VI (Petit *et al.*, 1987). Moyer and Josling (1990) attribute this difference to the following factors. The first and most important was the combination of personalities in 1987/88: Delors as Commission president and his vision for a SEM, alongside Christophersen and Andriessen, who were senior and successful commissioners. Their respective cabinets were able to rise above the daily administration of the CAP and impose a broader view of the situation. Moyer and Josling (1990, p. 86) wrote, 'The success of Delors, Andriessen and Christophersen in forming an inner circle reflects both the severity of the crisis and the leadership capacity of individuals.'

This inner circle helped speed the reform process within the Commission in the 6 months after COM (87) 100. The alliance proved sufficiently powerful that other Commissioners and their attendant national interests in agriculture did not have to be compensated in the bargaining process. The influence of personalities is also central to the MacSharry reform process. The development of a public choice approach to CAP reform in Chapter 5 raises the structure vs. agency debate. The debate concerns the relative influence of individuals and the structure in which they operate, on the policy process. The role of Commissioner MacSharry in the 1992 reforms is discussed in Chapter 6, while Chapter 9 discusses how this can be reconciled with the institutions framework set out in Chapter 5.

Moyer and Josling (1990) note that the decision to propose production stabilizers to the College of the Commission was made by eliminating other options. Price cuts were unacceptable to the German government; mandatory quotas likewise for the UK and the Netherlands; and set-aside was considered by the inner circle to be expensive and had not seemed to work in the USA. Further, the idea of budget stabilizers triggered by a financial overspend was considered politically unacceptable, so stabilizers triggered by production were the preferred option. These production stabilizers were also judged to have a further advantage: they could be presented to the different national constituencies as a variant on the guarantee thresholds which had operated since the mid-1980s, and therefore as not a radical, far-reaching reform.

Moyer and Josling (1990, p. 88) sum up this process: 'The Commission kept both governments and the farm lobby at arm's length from its planning process, but was ever mindful that its final proposal would have to achieve approval in the European Council.' The speed of the process in 1987 in the Commission was slowed by the need for agreement of the 1987/88 price package (something which was repeated in the MacSharry reforms).

At the end of July 1987 the College agreed COM (87) 410, *Review of Action Taken to Control the Agricultural Markets and Outlook for the Common Agricultural Policy* (Commission of the European Communities, 1987c). Although the document contained no official figures, it formed the basis of the full legislative (and quantitative) proposals to the CoAM in September 1987 (as discussed below). Included in the document were proposals for a new, fourth budgetary resource and a maximum guaranteed quantity (MGQ) for EU cereals production, any production beyond which would trigger unspecified penalties. A figure of 155 Mt was mentioned as the total annual level of EU cereals production before penalties would be triggered.

A significant difference in the circumstances of the enactment of the stabilizer regime compared with the milk quota reform was that the former was agreed as part of a wider budget agreement. This wider agreement was being sought by the Commission in an attempt to redress the institutional imbalance in CAP decision-making which allowed the CoAM to agree a level of agricultural spending which consistently exceeded any guideline set for it. The introduction of milk quotas in 1984 was accompanied by no such wide-ranging budget agreement and was much more a fire-fighting exercise to control the budget costs of the milk sector.

The Delors I Plan, promulgated in February 1987 and quoted in *AgraEurope* (26 June 1987, p. E/3), talked of the need for budget stabilizers, whether triggered by financial or production levels: 'Application of these principles would change significantly the present situation where agricultural regulations have direct repercussions in budget terms. Instead agricultural rules will be applied or amended in such a way as to ensure that pre-determined budget allocations are respected.' The automatic nature of stabilizers and their implication for the executive powers of the Commission were points of some contention during the reform process.

3.4.3 Reform Process: Phase 2

The full legislative proposals based on COM (87) 410 were presented to the CoAM on 22–23 September 1987. The stabilizer regime proposed for the cereals sector was as follows: an MGQ for total cereals production of 155 Mt (megatonnes), which if exceeded incurred the penalty of a combination of price cuts and increases in the CRL. This penalty was to be applied in 1% slices (1% increase in production equals a 1% increase in the penalties) with the first 1% exempt, up to a maximum penalty of 5%.

The initial debate in CoAM was characterized by consensus on the need to act to keep agricultural spending growth 'reasonable'. However, this term was sufficiently ambiguous that any consensus based on it was spurious. Agriculture ministers were split over whether budget or social priorities were the greatest considerations in reforming the CAP. The UK and Dutch representatives concentrated on the budget. West Germany was joined by the Mediterranean countries, Ireland, Luxembourg and Belgium in being staunchly against automatic price cuts. The French position, like that for the reforms of 1984 and 1992, reflected an ambivalence in its government's attitude to the purpose of the CAP.

The specific issues were a demand for an MGQ of 165 Mt by France and Germany, plus a German demand for a set-aside programme. Such a programme was opposed by the Commission, which argued that it would not be effective at controlling production. The situation in the CoAM during the last 3 months of 1987 was one of limited movement, partly because the budget problem was not yet at full crisis level (the EU would not run out of money for another 6 months) and partly because of a weak Danish presidency (Moyer and Josling, 1990).

This drift meant that agriculture was the outstanding element of the budget package considered by the European Council at Copenhagen on 5–6 December 1987. The summit collapsed when Chancellor Kohl refused to negotiate because the Council presidency's document contained no reference to set-aside. Further disagreement centred on how automatic and how big the penalties of the stabilizers should be.

3.4.4 Reform Process: Phase 3

The central feature of January and February 1988 was Germany's occupation of the presidency of the Council of Ministers. This led to a change in the German position. The priority of providing leadership of the EU to protect its credibility within the member states and in international negotiations competed with purely German farm interests. Moyer and Josling (1990) described the German farm minister, Keichle, as working diligently for acceptable compromise. A consensus was reached by the end of January. An MGQ of 160 Mt was agreed, the consistent German opposition to price cuts as a means of controlling budget expenditure was dropped to secure UK and Dutch support, and in return a set-aside programme was agreed. The penalty in the mechanism was to be the Commission's proposed mixture of price cuts and increases in the CRL. Further, the maximum penalty was reduced to 3% (against the Commission's proposed 5%).

The reform process of the stabilizer regime was able to be concluded when the West German government changed position. The West German position after 1985 had favoured some kind of change in the CAP and had concentrated on trying to influence that change. The Germans' objective for the CAP remained constant: the need to protect their small farmers. Their opposition to price cuts (especially when automatically imposed) remained through the period 1985–1988. When this opposition to price cuts was relaxed (and traded for other countries' compromises) a unanimous agreement was able to be reached in the European Council.

The 1988 reforms were agreed by unanimity because of their tie with the future budget arrangements of the EU. This was how Kohl was able to scuttle the Copenhagen summit. Moyer and Josling (1990) attribute the flexibility of the West German government's position in January 1988 to its occupancy of the presidency. Petit *et al.* (1987) make a similar attribution to the French government's position in the milk quota reforms. It is an important point because, as will emerge in Chapters 4 and 6, the

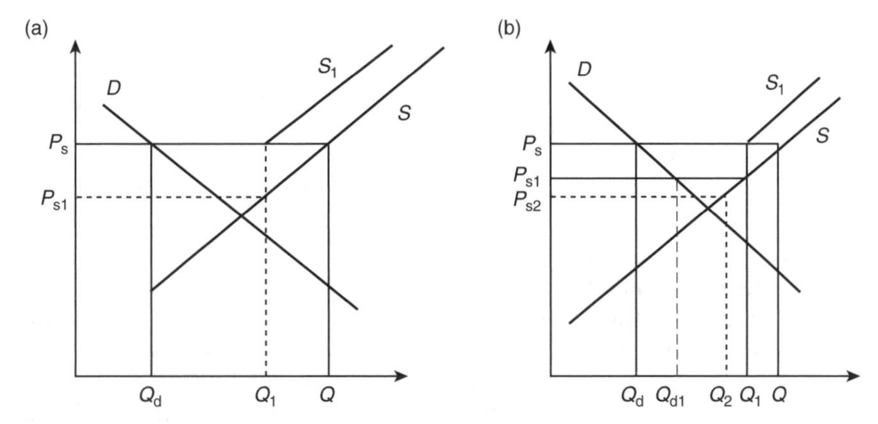

Fig. 3.1. The economics of the stabilizer regime. (a) The CRL; (b) the price cuts.

cumulative effect of the automatic price cuts which occurred under the stabilizer regime was an important factor in the enactment of the MacSharry reforms.

3.4.5 Economics of the Stabilizer Regime

Figure 3.1 illustrates the partial equilibrium effects of the additional CRL and lagged price cuts introduced in the stabilizer reforms. Prior to the reforms, output Q was produced in response to the support price P_s. The difference between total production and domestic demand at this price level, $Q–Q_d$, is exported at the world price. The introduction of an MGQ of Q_1 and a CRL of $P_s–P_{s1}$ shifts the supply curve for output above Q_1 to S_1, reduces output to Q_1 and exports to $Q_1–Q_d$. $P_{s1}–P_{s2}$ in Fig. 3.1b represents the lagged reduction in support prices. As a result of a $P_{s1}–P_{s2}$ fall in the intervention price, output falls further to Q_2. The effects of the lagged price reduction are similar to those of the increased CRL except that the lagged price reduction leads to a slight increase in domestic demand to Q_{d1}.

3.5 SIMILARITIES IN MILK QUOTAS AND STABILIZERS REFORM PROCESSES

This section briefly highlights the similarities in the policy processes associated with the CAP reforms of 1984 and 1988 respectively. It is the argument of Chapter 5 that these observations can be interpreted and organized using the common analytical framework. Chapter 8 considers whether the same framework can be applied to the MacSharry reforms.

The first similarity between the milk quota reforms and the stabilizer reform is that the debates in the CoAM which led to their enactment became serious only when a major budget crisis arose. In both cases the final agreement was very similar to the Commission's proposals, with only the numerical levels altered (in both cases made less tough/more generous by the CoAM). Even though the CoAM eventually accepted the type of CAP reform in the Commission's proposals and only changed the numbers, the reform process still took over a year and the near-bankruptcy of the EU to reach its conclusion.

The second similarity is that both reforms seem to display the characteristics of a fire-fighting exercise. The Commission's proposals focused on the milk sector in 1984 and the cereals sector in 1988. These sectors were at the time displaying the most rapid growth in expenditure.

Third, Moyer and Josling (1990) note that, except for the UK and Dutch representatives, each CoAM member was motivated almost exclusively by the farm interest when considering the reforms. Further, international pressure for reform of the CAP appeared inconsequential in the stabilizer and milk quota reform processes.

The fourth similarity is that the European Council provided an important role in both reform processes. In 1984, the conditional part of the CoAM's agreement at the meeting of 5–6 March illustrates an unwillingness to take a final decision to reform the CAP without the European Council having had an opportunity to reach agreement. The European Council was ceded the lead role in CAP reform; only when it failed to achieve agreement did the CoAM reach a reform agreement. The reforms agreed effectively resolved the few outstanding issues left after the European Council meeting of 19–20 March; a large part of the final agreement had already been informally agreed by the European Council (see Petit *et al.*, 1987). The European Council actually agreed to the 1988 reforms. As is described in Chapter 6, the European Council played a much less prominent role in the MacSharry reform process.

The reform processes in both 1984 and 1988 were concluded when the French and German governments' positions came together. Further, the reform negotiations in both cases were conducted on the basis of achieving unanimity. This ensured in both cases a complex package with balanced benefits and sacrifices between all member states.

3.6 CONCLUSIONS

This chapter has argued that a positive approach is required when trying to account for CAP reforms. Such an approach requires a full consideration of the political and the economic aspects of a reform process. A description of the political and economic circumstances surrounding the Mansholt Plan, the milk quota reforms and the introduction of the stabilizer regime has been provided. These accounts have shown that there is some degree of comparability between these significant attempts to reform the CAP before the MacSharry reforms of May 1992. There is a pattern to these reforms. Reforms are enacted only out of a sense of crisis. The CoAM is the central institution and enjoys a negative power in the reform process: it can delay or halt the reform process as a result of inactivity. This delay creates tensions with other ambitions of the EU (international trade negotiations or control of budgetary expenditure). This tension fuels the sense of crisis in which the European Council, as the highest political authority in the EU, gets involved. The reforms that are eventually enacted in response to this crisis are not substantially different from the initial proposals from the Commission. This is why the power of the CoAM in the CAP reform process is negative.

NOTE

1. The positive–normative distinction follows that made in the political economy literature. Stigler (1971) argued for the need for the economics discipline to move beyond the normative analysis of economic policies and prescriptions of which policies *ought* to be implemented, to a positive analysis of why certain policies *are* chosen.

Chapter 4

The Pressures for the MacSharry Reforms

4.1 INTRODUCTION

Chapter 3, by discussing the milk quota and stabilizer reforms, emphasizes the need for a positive approach to the CAP reform process. Such an approach would provide an account of the pressures which trigger the process, and the mechanisms which link those pressures to the eventual enactment of a CAP reform. This chapter examines the pressures which generated the environment for the MacSharry reform process.

A description of the circumstances, both economic and political, which were necessary for the MacSharry reform process is part of an understanding of that process. The environment within which the proposals for CAP reform were drafted and promulgated affected both the reform process and its outcome. This chapter is divided into three sections. The pressures for CAP reform are partly external, partly internally generated. The internal pressures, arising from the operation of the stabilizer regime, are covered in Section 4.2. External pressures, resulting from the effect of the CAP outside the EU, had two sources in the MacSharry reform process – the oilseeds dispute and the Uruguay Round – and these are dealt with in Sections 4.3 and 4.4 respectively.

4.2 THE OPERATION OF THE STABILIZER REGIME, 1988–1992

The basic details of the operation of the stabilizer regime are provided in this section. There is a brief survey of how the system operated between 1988 and 1992. Next, the histories of the price packages agreed in 1991/92 and 1992/93 are described. These coincide with the formal progress of the MacSharry reforms, described in Chapter 6.

4.2.1 The Stabilizer Mechanism: an Outline

The stabilizer regime was agreed for the cereals sector of the CAP in February 1988 by the European Council, as part of a wider budget agreement. This budget agreement set up guidelines for the next 4 years: the annual growth in the EAGGF spending was limited to 74% of the growth in EU GNP, starting from the ceiling for 1988 of ECU 27.5bn.

The stabilizer regime consisted of the following arrangements. The basic CRL of 3% of the intervention price (in place since 1977) was retained. The scheme introduced an MGQ of cereals production. This was set at an annual ceiling of 160 Mt for EU cereals production for the 4 marketing years after 1988/89. An additional or supplementary CRL of 3% was levied at the beginning of the marketing year. If cereals production exceeded the MGQ in any marketing year then two things happened: first, a levy proportionate to the amount of the overshoot and up to a maximum of 3% of the intervention price would be calculated. If this figure was equal to 3% then nothing would be returned; anything less than this 3% and the appropriate adjustments would be made for producers. The second effect of production exceeding the MGQ was an automatic 3% reduction in the intervention price for the next marketing year.

The February 1988 agreement also created a 'monetary reserve' account of ECU 1bn to mitigate the effect of ECU–$ exchange rate variations on the EU budget's export subsidy bill. This could be drawn upon if the exchange rate movement had adversely affected the budget to the extent of ECU 400m or greater.

4.2.2 A Survey of How the Stabilizer Regime Operated, 1988–1992

Table 4.1 presents the basic figures which describe how the stabilizer regime operated. The MGQ was exceeded in 3 of the 4 years before the MacSharry reforms were agreed. The additional CRL was waived in 1989/90 because the amount of overshoot was judged insubstantial (the automatic price reduction was triggered for 1990/91).

An important factor to examine with regard to a description of the MacSharry reform process was the harvest in 1990/91. This did not exceed the MGQ and therefore there was no automatic price cut in 1991/92 and

Table 4.1. Operation of the stabilizer regime, 1988–1992.

	1988/89	1989/90	1990/91	1991/92
Guideline (Mt)	160	160	160	160
Cereal production (Mt)	162.6	160.5	158.9	168.9
% overshoot	1.6	0.3	n/a	5.6
Cereal stocks (Mt)	9.9	11.7	18.7	26.4
Price cut triggered?	Yes	Yes	No	Yes

Table 4.2. Budget effects of the stabilizer regime, 1988–1992.

	1988	1989	1990	1991	1992
Guideline	27,500	28,624	30,630	32,511	35,039
Expenditure covered by guideline	26,400	24,406	25,069	30,961	35,039
% growth in expenditure	17.3	−7.5	2.7	23.5	13.2
Total expenditure	27,687	25,873	26,453	32,386	

Note: The figures for 1992 are deliberately provisional. They are the figures for the 1992 budget agreed in December 1991, which the CoAM had in May 1992 when it agreed the MacSharry reforms.

the additional CRL was returned. This was described as a 'fluke' by a senior official in DG VI, and nothing to do with the operation of the stabilizer regime controlling production. This affected the negotiation of the 1991/92 price package and progress of the MacSharry reforms. The MRQ was exceeded in 1991/92 and this affected the environment in which CAP reform was being negotiated. This difference is described below as the negotiations of the 1991/92 and 1992/93 price packages are separated. These dates are concurrent with the formal progress of the MacSharry reforms, described in Chapter 6.

As noted, the stabilizer regime was enacted as part of a wider EU budget deal in 1988. Its stated aim was to control the growth in EAGGF (Guarantee) expenditure. Any consideration of the operation of the stabilizer regime must include a brief look at its budget effects (Table 4.2). Further, as described in Chapter 2, most agricultural economics analyses of CAP reform centre on detailing the pressures for CAP reform, in particular the budget imperative as the cause of reform.

The success in controlling the growth of EAGGF in 1989 and 1990 was credited by a member of the MacSharry team to the strong dollar helping to reduce the export subsidy element of the EAGGF budget. It has been noted above that the penalties of the stabilizer mechanism were not triggered for 1991/92, but this period was characterized by a deteriorating budget situation, for reasons described in Section 4.2.3. Although the final expenditure figure quoted above was well within the guideline, considerable political effort was required in the agreement of the 1991/92 price package in order for this to be achieved.

4.2.3 The Price Package 1991/92

Ray MacSharry, the Agriculture Commissioner, originally intended that the proposals for the 1991/92 price package and proposals for CAP reform should be concurrent. The GATT negotiations were cited by the reform team as delaying the publication of the latter. COM (91) 100 (Commission of the European Communities, 1991a) was only a 'reflections paper' and not at that stage a formal proposal. The 1991/92 price proposals were introduced separately by the Commission. This separation was something that MacSharry and his reform team were keen to maintain throughout the negotiations which led to the 24 May and 18 June 1991 agreements of the prices for 1991/92.

The College of the Commission agreed the 1991/92 price proposals on 27 February 1991. The debate inside the Commission was heated, and the proposals agreed were very similar to those proposed by MacSharry and prepared by DG VI. The background against which they were agreed was a pending budget crisis. The latest DG VI estimate of EAGGF expenditure for 1991, in February 1991, was ECU 33bn. The 1991 budget (agreed in December 1990) had set EAGGF expenditure at ECU 31.6bn.

AgraEurope (editions late January and early February 1991) estimated that ECU 400m would be brought in from the monetary reserve account because of the effect of the fall in the value of the dollar against the ECU. This left a supplementary budget requirement of ECU 995m for the estimated 1991 EAGGF expenditure to be funded. Depending on the use of the monetary reserve account (which was budgeted outside the guideline) and certain flexibilities in the calculation of the guideline, *AgraEurope* (in its editions of March 1991) predicted that total EAGGF expenditure would exceed the guideline for the first time since the 1988 budget agreement.

The option of solving the budget crisis by raising the guideline was vehemently opposed by MacSharry. The MacSharry cabinet believed that any relaxation of the guideline would have compromised the impetus for fundamental CAP reform. The raising of the ceiling on EAGGF expenditure was supported by Jacques Delors. This was one of the low points of a difficult relationship between MacSharry and Delors. Grant (1995b) provides an account of this dispute. Delors wanted the College to agree an increase of ECU 1.3bn in the budget guideline, excused by the exceptional nature of German unification. The College agreed with MacSharry's opposition to raising the guideline and agreed price proposals which kept predicted spending within the guideline. After an exchange between the two about whether Delors would defend the position in public, communications were broken off for a period.

The price package agreed by the College was very similar to that proposed by DG VI and MacSharry's team. The sectors targeted were beef, milk and tobacco, which were diverging by 100%, 31% and 6.5% respectively from their 1991 budget estimates. The following proposals were agreed. Price cuts of 15% in the intervention price for tobacco, 5% for sugar, 7% for durum wheat and 3% for rice were agreed. Further, the beef intervention regime ceilings were tightened and the global milk quota cut by 2%. With respect to the stabilizer regime, a doubling of the basic CRL (to 6%) was proposed, along with a special 1-year set-aside scheme.

These proposals were first presented to the CoAM on 4 March 1991. The

Councils of 4 March and 25–26 March were notable for what MacSharry's team termed a 'strong' warning from the Agriculture Commissioner that the 1991 guideline was about to be breached. Only the UK and the Netherlands spoke in favour of respecting the guideline and discussing the price package as proposed. All of the other members spoke against the severity of the Commission's price proposals and argued in favour of raising the guideline.

MacSharry stated that if the Commission had proposed raising the guideline this would have required unanimity in the CoAM and/or the European Council. John Gummer, the UK Secretary of State for Agriculture, was definite in stating that he would veto any move to raise the guideline for 1991. There was a similar ten–two majority in favour of raising the guideline in Ecofin (the council of finance ministers of the member states) as well. The reason cited by the majority of these ten was that German unification represented an 'exception' provided for in the 1988 budget deal.

The 22 April meeting of the Council of Agriculture Ministers failed to agree a 1991/92 price package and the ten–two majority in favour of raising the guideline persisted. *AgraEurope* 27 April 1991 reported talk of a softer package and accounting gymnastics, based on the acceptance of the 1991 guideline of ECU 32.5bn as an impassable constraint in the agreement of a price package.

The worry of the MacSharry team was that a deadlock in the price package negotiations would mean that the issue was forced on to the agenda of the European Council meeting at the end of June (as had happened in previous budget crises in 1984 and 1988). The convention of price reviews was that agreement was reached by the end of April. (This deadline was exceeded in the mid-1980s; the record stands at 17 July 1988.) A European Council agreement on the farm budget and its future constraints would pre-empt the MacSharry reform proposals. In particular, it might produce CAP reforms of the type of 1984 and 1988, which MacSharry believed had produced short-term solutions to immediate fiscal problems, whereas the proposals of COM (91) 100 augured a new direction for the CAP.

Agreement was partially reached on 24 May 1991. The prices for the cereals regime were agreed at the subsequent CoAM meeting, on 18 June. The bare details were that institutional prices were rolled over from 1990/91 for most products. The basic CRL was raised from 3 to 5%. A 1-year special set-aside scheme was agreed, plus a 2% cut in milk quotas, a 1.5% cut in oilseed and protein crop prices and a 2% cut in basic sheepmeat prices. At the 18 June CoAM meeting a 3.5% reduction in intervention price for durum wheat was agreed with the 1990/91 intervention prices for common wheat, barley, maize and other cereal crops rolled over to the next year.

The reaction of *AgraEurope* was to rail against what it considered a classic budgetary 'smoke and mirrors' agreement. The CoAM agreement would bring 1991 EAGGF spending to a level well below the guideline.

This was achieved through a number of reworkings of the original budget. The most important of these was the extrapolation of the benefits of a recent strength in the dollar on the export subsidy bill through the next 2 financial years; this took ECU 300m off the 1991 budget, and between ECU 1bn and 2bn off the preliminary draft for the 1992 budget.

The rejection of the Commission's proposed 5% cut in sugar prices meant that ECU 300–400m could be saved from the 1991 budget, because the import threshold and thus import levy revenue would be higher. *AgraEurope* (31 May 1991) also cites optimistic calculations in the reworked 1991 farm budget of the effects of the milk quota cuts and tightening in the operation of the beef intervention regime.

Members of the MacSharry cabinet consider that the rise in the value of the dollar in early summer 1991 was most helpful in expediting the 1991/92 price package agreement.

4.2.4 The 1992/93 Price Package

Chapter 6 details how the need to agree the 1992/93 price package was the key factor in the conclusion of the MacSharry reform process at the May 1992 CoAM session. This section notes that it was the toughness of the operation of the stabilizer regime which provided the backdrop to the debate on the 1992/93 price package and ultimately the MacSharry reforms. This toughness was the series of automatic price cuts detailed in Table 4.1. The toughness is in terms of the view of a qualified majority of the CoAM rather than agricultural economists.

The difference between the 1991/92 price package and the 1992/93 price package was the context of automatic price cuts implied by the operation of the stabilizer regime in 1992/93 but not 1991/92. In a sense, the effectiveness of the stabilizer mechanism in controlling and cutting the level of nominal support prices was one of the key factors prompting the MacSharry reforms.

4.3 THE OILSEEDS DISPUTE

The oilseeds dispute, which persisted between 1985 and 1993, warrants its own section as a cause of the MacSharry reform process for a number of reasons. There is the general-interest point that it provides an illustration of how neutered the General Agreement on Tariffs and Trade (GATT) was in its role of world trade policeman, and demonstrates one of the challenges that faces the World Trade Organization (WTO). More specific to this work, the oilseeds regime enacted in November 1991 (prior to the MacSharry reforms) affected the decision to reform the cereals regime in May 1992. The extent of this influence is described in this section and developed further in Chapter 6.

The decline in the competitive position of the EU cereals sector was given by MacSharry as one of the concerns he had for European agriculture during his time as Commissioner. The French government also lobbied very hard on this issue. The specific concern was the growth in the amount of cereal substitutes entering the EU from the USA under very low rates of duty. The Dillon Round had allowed energy and protein crops to enter at zero tariff. Manioc was energy and soya was protein. Also, maize gluten feed, a by-product of the production of ethanol that could be fed to cattle, was again imported at a zero rate of duty.

MacSharry brought this concern to his agenda for CAP reform and the Uruguay Round negotiations. It is in the oilseeds dispute that the link between international pressure and CAP reform is most clearly observed. Oilseeds were already under the auspices of the GATT, but, as discussed below, the Uruguay Round and the oilseeds dispute became intertwined.

The link between international negotiations in the Uruguay Round and the oilseeds, and the various changes in the oilseeds support regime by the EU in the period described is not straightforward. This section is an attempt to develop a background to allow some kind of examination of its role as a causal factor in the MacSharry reform process. Section 4.3.1 provides a background to the development of an oilseeds industry in the EU and the contribution of the support system to that development. Sections 4.3.2–4.3.4 look at the oilseeds dispute before the enactment of the MacSharry reforms. This chapter considers the oilseeds dispute as a possible cause of the MacSharry reforms and therefore stops the account of the oilseeds dispute at the point when the reforms occurred. The effect of the enactment of the MacSharry reforms on the resolution of the oilseeds dispute is detailed in Chapter 7.

4.3.1 Background to the Oilseeds Dispute

The oilseeds industry comprises three parts: the harvesting of seed, the processing of cake and the production of oil. The background aggregate data are usually presented in terms of three units: seed equivalent, cake equivalent, oil equivalent. The industry has growers, feed compounders and crushers. The latter two are grouped as processors.

The growth of a significant EU oilseeds industry since the 1960s had created USA resentment. EU production of oilseeds ballooned from 600,000 t in 1966 to over 12 Mt in 1990 (the middle of the oilseeds dispute). These figures are from various issues of *Agricultural Situation in the Community* and are in oil equivalent. This resentment was exacerbated in the 1980s. The USA Department of Agriculture presented the argument that EU imports from the USA reached $3bn in the early 1980s but less than $2bn by 1988. Over the same period EU output rose from 2.5 Mt to 10.3 Mt (seed equivalent). Indeed, in the period 1985–1992, during which the oilseeds dispute took place, production grew further to close to 12.5 Mt (of oilseed equivalent) in 1990.

The USA resentment was derived from the perception that a competitor to the USA oilseeds industry had been created by unfair government support. EU farmers were guaranteed two to three times the world price for rapeseed, sunflower seed, soybeans, beans, peas and lupins. The specific charge made by the USA was that processors were given direct aid so that EU oilseeds were bought, rather than imported ones.

This resentment took legal form in a complaint to the GATT panel in 1985. Importantly, the oilseeds industry was already under the auspices of the GATT after the Dillon Round of 1960–1962. Under AII of GATT (the original 1947 agreement), which provides for schedules of concessions to be agreed, the EU had a standing commitment from 1962 to charge zero duty on imported oilseeds. Thus, unlike all the rest of the CAP regimes, the oilseeds regime has always been under the auspices of the GATT. As can be noted from the figures above, this was initially no great concession on the part of the EU; the industry had been tiny.

The EU, when establishing a support regime for oilseeds in 1966, had worked around this constraint (as part of the construction of the CAP). There would be a guaranteed price for growers varying from two to three times the world average. A direct income payment (DIP) would be made to the processors on the basis of the difference between the world price and the Community guaranteed price which the processor had to pay for oilseeds of EU origin. Theoretically (at least in bureaucratic terms), the purchase of EU oilseeds was subsidized to a level where the net price to the processor was the same as the world price. However, it was the operation of this system which provided the greatest grist to the USA mill and led to the referral to the GATT Panel.

4.3.2 The Oilseeds Dispute Prior to the MacSharry Reforms

In December 1989, a GATT panel found the EU oilseeds subsidy regime in violation of the basic principles of the GATT (laid down in the original 1947 agreement). The panel found the EU 'guilty' on two counts. Count 1 was a violation of AIII.4 of the GATT by the processor subsidy scheme, which favoured domestically produced oilseeds over imported ones. The operation of this scheme is described below. Count 2 is that the same processor subsidy scheme contravened the 1962 commitment to charge zero duty on imported oilseeds.

Article III.4 of the original 1947 GATT agreement deals with ensuring that there is no national bias to any regulation or taxation. Specifically, it mentions the 'regulation of the mixing, processing or use of a product'. This means that there can be no amount guaranteed for domestic producers, and that a contracting party to the GATT cannot restrict the 'mixing, processing or use of a product' in order to protect a small-scale national industry in that product.

The panel found that the operation of the EU's oilseeds regime contravened the Article in the following sense. The world price used to calculate the subsidy for the processors was a *reconstructed* world price and could in certain circumstances diverge from the actual world price. Hence the possibility existed of compensation greater or less than the difference between the EU guaranteed price and the price that was trading on the world market. Further, under the EU scheme processors did not have to provide any guarantees that the price they paid was equal to the EU guaranteed price. If processors could manage to get them cheaper then, with the subsidy, the EU oilseeds were effectively cheaper than those on the world market.

The regime was sufficiently open to the possibility of overcompensation and discrimination for the Panel to feel able to rule against the EU under AIII.4. This also extends to count 2, where this discrimination was ruled to nullify the legitimate expectations of the 1962 deal, namely that the USA would enjoy free trade and allow its competitive advantage in the production of oilseeds and would be free to exploit it. The ruling called on the EU to adjust the regime, but it was given a 'reasonable' period in which to do so.

4.3.3 Response 1: the Rebalancing Option

The Commission acquiesced to most of the report. There was the obligatory stab back that the USA was guilty of the same practices in its dairy and sugar sectors (neither of which was covered by GATT). The expectation was that the issue would be thrown into the Uruguay Round for settlement with the rest of agriculture by December 1990 at Brussels. This was the genesis of the EU's proposal for rebalancing: that reductions in trade-distorting farm subsidies should be measured against an aggregate measure of support (AMS) or *global* yardstick, not on a commodity-by-commodity basis. In this way, oilseeds could be brought back into line with the rest of the arable crop sector of the CAP, with the 1962 pledge in effect being waived.

The EU offers in the Uruguay Round before the collapse at Brussels focused on the rebalancing option – all this based on the idea that the EU could escape the GATT negotiations by offering a reduction in some measure of aggregate EU farm support in exchange for protection from grain substitutes. However, this response was premissed on a successful conclusion to the Uruguay Round by December 1990. This did not happen; thus another response was required by the EU.

4.3.4 Response 2: a New Oilseeds Regime

This second response was to reform the oilseeds regime to the satisfaction of the original ruling. After much debate and compromise, the CoAM agreed a new oilseeds regime on 22 October 1991 (2 years after the original ruling). This was, according to the Commission, 'a compensatory payment system with per hectare aids paid direct to the producers'. Direct compensation was paid to the grower for selling his or her oilseeds at the unsupported world price. This replaced the processor subsidy of the old regime. The guaranteed price level in the EU was removed, with the processor able to choose between EU and imported oilseeds, which should trade at similar prices.

The compensation was area-based. Each farm had a defined maximum guaranteed area (MGA) equal to the current area planted of the three oilseeds (rapeseed, sunflower seed and soybean). Compensation was based on, and limited to, this MGA. The compensation was paid in two parts. Producers automatically receive

$$(\text{Target} - \text{world price}) \times \text{average yield} \times \text{MGA}$$

The average yield figure could be adjusted up or down on a regional basis; each member state had the choice of using the oilseed or cereals yields for each region – subject to overall compensation for that member state not being greater than if oilseed or cereal yields had been used in all regions of that state. The choice of which yield to use was designed to avoid unbalancing the relationship between rape and sunflower seed production (by trying to provide two different compensation schemes for the two products). The second payment was equal to the difference between the automatic payment and the full payment due on the basis of the actual *observed* reference price. The world price, labelled above, was an initial *projected* reference price.

The compromise paper required at the CoAM on 21–22 October to reach agreement on this system made it clear that any agreement on oilseeds did not in any way affect the CAP reform debate. The new regime was agreed to only apply to rapeseed, sunflower seed and soybeans harvested in 1992. The scheme was enacted only for 1 year, because plans for an oilseeds regime already existed in COM (91) 258 (the MacSharry reform proposals; Commission of the European Communities, 1991b). This oilseeds regime was a temporary measure to meet the immediate demands of the GATT ruling. The longer-term objective of the MacSharry reform team was to enact an oilseeds support system as part of the MacSharry reforms, which was similar to that for cereals and from which a GATT position similar to that for cereals could be negotiated (ending the 'special' status of oilseeds described in Section 4.3.1).

Even though the CoAM communiqué which accompanied the agreement of the 1-year oilseeds scheme stressed that there was no link with CAP reform, a mechanism discussed in Chapter 6 is the psychology of an agreement on a major sector of the CAP, even for a year, which moves the CAP in a direction similar to that proposed in COM (91) 258. This direction is towards area-based direct income compensatory payments, although the system itself was effectively a deficiency payments system. The scheme infuriated the USA, which again referred the oilseeds regime of the EU to the GATT.

In April 1992 the EU oilseeds regime of October 1991 was ruled to involve 'a systematic offsetting of the effect of the general movement of import prices on production levels'. Producers were directly compensated for any price advantage imported oilseeds might enjoy. This compensation was product specific and directly linked to current production levels. This meant that it contravened the tariff concession of the original 1962 agreement. The second compensation insulated producers from movements in import prices, thus denying the possibility of imports enjoying some price advantage. This time the panel called upon the EU to implement the ruling without delay, unlike in the first case.

The MacSharry reforms were the response to the second GATT panel ruling on the oilseeds regime. They brought oilseeds into line with the arable crop regime (i.e. DIPs and set-aside – see Chapter 6). The whole of the MacSharry reforms were then part of the negotiations in the agriculture section of the Uruguay Round. Chapter 6 discusses political leadership in terms of MacSharry's attempt to get the oilseeds dispute settled as part of the wider Uruguay Round agreement (when in fact it was a dispute distinct from the Uruguay Round). A reduction in the vulnerability of the European cereals sector to import substitutes was always an objective of the MacSharry reform team. The US negotiators were faced with the following options. First, they could reject the MacSharry reforms as an acceptable system of support in the GATT and risk jeopardizing the entire Uruguay Round; or second, they could accept the MacSharry reforms as part of a post-Uruguay Round GATT. The latter option contained the additional link that once the MacSharry reforms of the EU arable sector were accepted under new rules (those agreed in the Uruguay Round), then international politics would not allow the oilseeds regime to be condemned under old rules. Once the USA agreed that the MacSharry-reformed CAP was acceptable in the GATT, then the EU's 1962 pledge was effectively rescinded.

The oilseeds regime was thus brought into the Uruguay Round negotiations (despite the fact it was already under the auspices of GATT and had twice been ruled against). This aroused American ire. Carla Hills, the United States' Trade Representative, threatened punitive action if the EU did not reform its oilseeds regime in a way that obviously was in accordance with the GATT panel ruling; the MacSharry reforms on oilseeds did not meet this criterion. Hills argued that the whole credibility of GATT as a multilateral trade organization was at stake despite the USA's trying to lever agreement using unilateral sanctions.

4.4 THE URUGUAY ROUND, 1986–1990

The negotiations of the agriculture section of the Uruguay Round were close to the MacSharry reform process both in their timing and in the substance of what was being discussed. Hence it is an obvious thesis that the Uruguay Round at different times was a causal factor in the MacSharry reform process. In order to describe such influence this section concentrates on the periods where the CAP decision-making system was directly affected by the demands of the Uruguay Round negotiations; that is, when the EU had to present a position, respond to a position or prepare a negotiating mandate. Three times the agriculture section of the Uruguay Round entered the politics of the EU from the realm of international negotiation: in 1990 and the run-up to the initial date set for the conclusion of the round; on publication of the Draft Final Act to try to restart the collapsed round; and, after the enactment of the MacSharry reforms, the agreement at Blair House in December 1992 and its subsequent ratification by the CoAM. This chapter looks at the causes of the MacSharry reforms, so Section 4.4 comprises subsections which describe the first two occasions that the imperative of a GATT agriculture agreement affected the CAP decision-making system. The effect of the reforms on the conclusion of the agriculture part of the Uruguay Round is considered in Chapter 7.

The description of the Uruguay Round as a causal factor concentrates on the bilateral negotiations between the EU and the USA on agriculture. The Cairns Group had some influence in the negotiations, but it was the bilateral relationship between the EU and the USA which was the key to any agreement on agriculture being reached in the Uruguay Round.

The background question is why these two participants were playing this game of negotiating reductions in agricultural support levels. It is not sufficient to take an economist's view of patterns of global trade and make inferences about the need for an extension of the GATT. To understand why these two entities agreed to discuss agricultural trade liberalization over 7 years requires detailed analysis of interests and institutions on both sides. This is a complex task beyond the scope of this book. The question with regard to the explanation of the MacSharry reforms is: if the demands of the Uruguay Round were at least one causal factor in bringing about the reforms, why did the EU allow agriculture to be included as part of the round in 1986? Two answers can be provided for this question. The first is that in previous GATT rounds the USA had signed the final agreement without any genuine progress on international agreements affecting domestic agriculture support systems. The EU was playing this game because of the belief in the Commission and member states that this pattern would repeat itself.

The alternative view is that the Commission and the member states wished to preserve the GATT or some kind of multilateral international trading system and also enjoy the benefits of the Uruguay Round. They believed that the USA would participate in both these wishes only if agriculture was included in the negotiations, hence the Commission and the member states were willing to consider international constraints on the CAP.

The background to world trade in agriculture in the 1980s is instructive for a consideration of the positions of both the USA and the EU. Between 1980 and 1989 the volume of world agriculture trade grew by 26% (roughly a third of the figure for manufactured goods trade); for

1970–1979 the comparable figure for agriculture was 54% (roughly equal to trading in manufactured goods). Over half of the growth in agricultural trade in the 1980s was internal to the EU, and therefore to an extent stimulated by the CAP. Agricultural production in the EU soared, and through subsidized exports had substantially eroded USA market shares. Farm exports as a percentage of total USA exports had fallen from 25% in 1974 to 11% by 1990 (figures from *The Economist*, 6 October 1990). The recovery of market share in world agricultural trade was one reason for the USA to push for agricultural trade liberalization.

Both the EU and the USA faced budget difficulties against a backdrop of the rising cost of agricultural support. The producer subsidy equivalent (PSE), according to OECD figures, grew in the crop sector from 45% to 66% in the EU, and from 8% to 45% in the USA, between 1979 and 1986 (Ingersent *et al.*, 1993, p. 80). The EU was facing the budget squeeze which led to the establishment of the stabilizer regime in February 1988 (Section 3.4). Similarly, around 1985 to 1986 was the height of the 1980s growth in the USA Federal budget deficit and the time of the Gramm–Rudman–Hollings Act of deficit reduction measures. Thus both sides had an interest in controlling the growth in agricultural support spending.

It is possible that the interests of each side evolved and changed during the 7 years of the Uruguay Round. The period 1989–90 was marked by a severe drought in the US Midwest. World cereal prices rose and the cost of agricultural support fell in both the EU and the USA, hence one pressure for negotiated reductions in support was mollified. The recovery of market share objective existed for the USA throughout the Uruguay Round.

4.4.1 Punta del Este Declaration

The Punta del Este declaration of September 1986 committed the contracting parties to negotiate a 'greater liberalization of trade in agriculture' and 'more operationally effective GATT rules and disciplines' regarding 'all measures affecting import access and export competition'. The commitments made were no more specific than that there would be an 'increasing discipline' of domestic agricultural subsidy. However, the fact that the reform of domestic agricultural subsidy regimes was on the agenda represented a break from the way agricultural negotiations had gone in previous GATT rounds. In the Kennedy Round (1964–1967) and the Tokyo Round (1973–1979) the EU had gravitated to a position of proposing concerted market-sharing agreements, but being explicit in the principle that the CAP would not be the subject of international trade agreement. It seemed that the Punta del Este declaration changed that position. The progress of the subsequent negotiations, however, could support the argument that the view of the EU on the relationship between the GATT and the CAP remained unchanged.

The negotiations for the Uruguay Round began in February 1987 and by the end of 1988 each side had tabled their opening negotiating positions. The USA called for the elimination of all forms of trade-distorting farm subsidies over 10 years. Only fully decoupled DIPs (i.e. those not linked to production in any way) would be allowed. This became known as the zero option, and much of the history of the first part of the Uruguay Round centres on the EU's reaction to it. The EU's initial response to this pre-emption was to emphasize short-term reform. Proposals were made for the reduction of instability and volatility on world commodity markets by international market management. This included plans to reduce world surpluses by a series of concerted actions for problem commodities: sugar, dairy, cereals, rice, oilseeds and beef.

In October 1988 the EU proposed an unquantified reduction in support over 5 years from a 1984 base. Any reduction in agricultural support would have to be measured by some kind of AMS which gave credit for domestic supply control independent of price and exchange rate fluctuations, and was also flexible enough to allow some increases in sector-specific supports. The EU introduced the support measurement unit (SMU) in this October 1988 proposal. Further, this proposal contained the explicit point that variable import levies and export subsidies were inviolable in any negotiations of agricultural subsidy reduction. Thus, despite the words of Punta del Este, the EU was conducting the agricultural negotiations no differently from the Tokyo or Kennedy Rounds. However, the EU seemed to have misjudged the intentions of the USA which were very different from in those previous rounds.

4.4.2 Mid-term Review, December 1988

The GATT contracting parties met in Montreal in December 1988 to review progress across the whole Uruguay Round. Agriculture, known as NG5 (negotiating group 5), was in deadlock, as suggested by the separation of the two opening proposals described above. Four problem areas – intellectual property, textiles, emergency import restriction measures and agriculture – were adjourned to give time for the GATT Secretary-General to sketch a compromise.

At this time a series of summit meetings between Yeutter (USA Agriculture Secretary) and Ray MacSharry took place in which the EU regarded itself as trying to secure 'a major tactical advantage' over the Americans (*Financial Times* (FT), 16 March 1989) by focusing on short-term demands. This included pressing for immediate reductions in target prices and a reversal of the recent relaxation of the USA set-aside programme. Further, the EU was pressing hard for a flexible freeze in farm subsidies in 1990 ahead of expected medium-term reform. The tactic was clearly a counter to the USA position of holding to the zero option proposal.

During this period the first reports emerged of a split in the EU camp between Andriessen (External Affairs Commissioner) and MacSharry. The former is quoted in the FT (30 January 1989) as saying that agriculture is 'not so important that the whole GATT Round should fail. That price would be too high.' MacSharry was much less enamoured with the politics of keeping the USA happy to secure a wider agreement in the Uruguay Round.

4.4.3 'Geneva' Compromise, April 1989

Arthur Dunkel (GATT Secretary-General) got the negotiations under way after the mid-term hiatus with a compromise paper that stated that agricultural negotiations should proceed on the basis of 'substantial progressive reductions in agricultural support and protection sustained over an agreed period of time' rather than the 'elimination' of the USA's zero option. Agreement was made to freeze the level of farm support immediately until the final agreement in the Uruguay Round, which was expected at Brussels in December 1990. The EU acceded to 1986 rather than 1984 as the base year for the measurement of support reduction. The main intention of the Dunkel draft was to bring about a dialogue and a gradual closing of positions on the agriculture issue over the 18 months before Brussels. The main groups were required under the Geneva Accord to submit proposals for long-term reform. MacSharry believed victory had been achieved by the EU – 'crowing' was one description (FT, 11 April 1989). The concession of a 1986 start date still meant that the EU had considerable 'credit' already on the reduction of farm subsidies.

The USA presented its submission on long-term reform as required by the Geneva Accord in October 1989. This mapped out three areas: (i) internal support; (ii) market access; and (iii) export subsidies. Under heading (i) internal supports would be divided into *red, amber* and *green* boxes. Red-box supports are those that were definitely trade-distorting and would be completely phased out over 10 years (this was close to the zero option supposedly abandoned at Geneva, and this infuriated the EU). Amber-box supports would be monitored and subject to GATT discipline. Green-box supports were those supports judged to create no trade distortions, and they would be allowed to continue. Under heading (ii) the USA proposed that all import barriers not explicitly allowed under GATT rules would be eliminated. Those 'permitted' non-tariff barriers (NTBs) would continue subject to conversion to tariff quotas in the short run (phased out after 10 years) and bound simple tariffs in the long run. The export subsidy heading had a proposal for the complete phasing out of these subsidies in 5 years.

In the 'propaganda war' outlined later, this set of proposals had two distinct advantages. The USA was explicit in stating that this meant the end of its Export Enhancement Program (EEP)[1] and the end of its deficiency payments system. If agreed by the EU, it would have meant the end of VILs and VES and thus the CAP as it was then operating. The fact that the USA was proposing an overhaul of its own agricultural support system as part of international trade negotiations caused discomfort to the EU. This was perhaps a lesson learned for the MacSharry plan for CAP reform. MacSharry was forced on to the defensive. It was his judgement that the CAP was more vulnerable in the Uruguay Round than the USA support systems. Accusations were made that the USA had reneged on the Geneva agreement. An FT leader (6 November 1989) called the proposal 'arguably even more draconian than its stand at Montreal a year ago'.

The EU position was delivered late in December 1989. The proposal was very general and contained no substantive figures to precipitate negotiation. It was a commitment to a gradual reduction in internal supports over 5 years. This *implied* reductions in the other two fields identified in the USA offer. The position was explicit in rejecting USA demands for specific commitments under the USA's three headings. There was also some hint of partial tariffication. At least part of the VIL would be fixed and subject to reduction as a *quid pro quo* for rebalancing. There would still be a variable element to reflect currency and 'other' fluctuations. Rebalancing was the EU's proposal, which allowed the raising of tariffs in some sectors against an overall or total fall in support.

Guy Legras (head of DG VI) was voluble at this time in citing the traditional argument against too market-based a regime: price volatility creates income fluctuations. The talks again reached impasse. The USA would not accept the EU's unwillingness to make commitments in any area other than internal supports (the USA was especially keen on export subsidies) or its insistence on rebalancing, which it regarded as a tactic for introducing tariffs on oilseeds. Section 4.3 describes this separate dispute, which was already covered under GATT rules. The 15 negotiating groups of the Uruguay Round were to produce 'framework' agreements by mid-1990 in order to give 6 months for the final details to be worked out.

Early in 1990, MacSharry was angry about the USA Farm Bill on two counts: the proposal to extend the EEP, and the part of the bill which allowed farmers to keep DIPs if they switched production to oilseeds. Further, there was no limit put on the budget allocation for the bill. Yeutter for his part denied that the bill was intended as a lever in the GATT negotiations, but he was willing to make a clear link between domestic policy reform and ongoing international negotiations. Yeutter stated that 'it would be foolish to say that USA Farm legislation will not be affected by the GATT negotiations or vice versa' (*AgraEurope* Green supplement, February 1990). The 1990 USA Farm Bill was described at the time in press commentaries as relatively aggressive (see *AgraEurope* editions at that time). The further trouble was that it encouraged an EU view that the USA was 'bluffing' in its strident demands for agricultural support reductions. The calculation was as follows: by the time the Uruguay Round would be entering its final stage, this bill would have become law and the USA would have changed its system of support in a way that was nothing like the substance or language of the zero option proposal in the Uruguay Round. It was not credible to believe that the US Congress would immediately rescind the legislation. The Bush administration would have a major problem getting the Senate to ratify anything subsequently agreed in the Uruguay Round which changed this Farm Bill. Hence the EU would be negotiating with the USA in the final stages of the Uruguay Round on the basis of the 1990 Farm Bill.

The Andriessen–MacSharry split was again in evidence at this time. In an interview with the FT (6 April 1990) Andriessen stated that in practice agricultural matters were dealt with 'very largely' by the Agriculture Commissioner, but the overall responsibility for the round was his and he would not allow the discharge of that function to be affected by adverse progress in one sector.

On 31 May 1990, an OECD meeting broke up in public disagreement with the usual diplomatic niceties not being observed. The USA had again pushed the programme of October 1989 and levelled criticism at the EU stance. MacSharry went on the offensive. A total elimination of farm subsidies would mean EU farm prices dropping by between 20% and 35% and 2–3 million farmers being forced off the land. MacSharry referred to an informal trade ministers' meeting in Mexico in April that separated farm discussion under the three headings of the USA, and which Andriessen had attended. MacSharry rejected such an approach because it could mean that the EU negotiated away its export subsidies while other countries kept other kinds of subsidies. MacSharry was gaining a reputation on both sides as hardline; he was quoted in the FT (1 June 1990) as saying, 'We haven't come up the river in a bucket.'

4.4.4 De Zeeuw Draft Paper, 27 June 1990

The De Zeeuw paper was published as the basis for a compromise at the G7 Summit at Houston between 8 and 11 July. At the summit, the rumbling agricultural trade negotiations dominated proceedings ahead of aid to the Soviet Union; the issue of CAP reform forced itself above the endgame to the Cold War. Some agreement was reached by the leaders on the basis of De Zeeuw. The final communiqué reiterated the intention to agree 'substantial progressive reductions' in agricultural support (in line with Geneva) using 'a common measure of support', and commended the De Zeeuw paper 'as a means of intensifying negotiations'. This was the outcome of intervention at the highest political level. However, the EU was in no way committed: only four of its leaders were there and the final communiqué only 'commends' the De Zeeuw paper. The USA and the Cairns Group again seized the initiative by adopting De Zeeuw as the vehicle for compromise. Two deadlines were agreed at Houston. By 1 October all contracting parties were required to submit to the GATT Secretariat an estimate of their current level of agricultural support (measured using the OECD's PSE), and by 15 October details of the reductions they were prepared to offer and over what time period.

The De Zeeuw paper was a middle position between the EU and the USA. It was closer to the USA on the export subsidy issue (the paper was described as 'least compromising' on the point that export subsidies should be reduced), but closer to the EU on the mechanics of internal support reduction: 1988 would be the base year and the EU's AMS would be the measurement tool. The actual question of how much and how quickly was left for the substance of the negotiations – a clear indication of how far behind the talks were in terms of the December deadline. The De Zeeuw paper was in no way the framework for the final agreement; it was still mainly about procedure.

The export subsidy issue was the crux of the problem for the EU. It highlighted the Andriessen–MacSharry battle. MacSharry went public with new figures that the end of export subsidies would mean the end of the CAP, and 3–4 million farmers coming off the land. MacSharry was much more robust and open about his interest in defending the CAP than he had been previously. For example, his public comments were about the CAP as 'social cement' and one of the 'central pillars' of the Community. The export subsidy issue was the one chosen by the USA to increase the pressure on the EU. It had appeal to the Cairns Group and was used by the USA to try to isolate the EU.

4.4.5 The Agreement of the EU Position Ahead of the Intended Conclusion of the Uruguay Round at Brussels in December 1990

MacSharry presented proposals to the College of the Commission for the final negotiating position of the EU in the agriculture section of the Uruguay Round on 19 September 1990. The proposal contained three main elements. The first was a reduction in internal support for agriculture by 30% over the period 1986–1995. The date 1986 was chosen so that the EU could enjoy the advantage of the reduction in measured support which had occurred since the Punta del Este declaration. The base from which the reduction would be taken was an AMS calculated on the average support levels between 1986 and 1988. The second main element was the tariffication of non-tariff import barriers. An element of this total tariff would be defined as fixed and reduced by 30% over the period. The calculation of the fixed element would be made on the average prices (world and domestic) for 1986–1988 also. There would be a concomitant 30% reduction in export subsidies. The rebalancing proposal of December 1989 was retained.

The proposal was rejected at the 19 September meeting. Frans Andriessen (the Netherlands), Martin Bangemann (Germany) and Sir Leon Brittan (UK) voiced the criticism that the 30% cut over 1986–1995 was too small and the rebalancing proposals were excessive. A campaign by MacSharry and his team and the intervention of Jacques Delors brought the College to agreement on 3 October (something he was not consistent about doing through the Uruguay Round; see Grant (1995) and the details in Chapter 7 of the Blair House Accord). The package was the same as before, except for a slight alteration in the wording, to allow Andriessen greater flexibility over the level of protection and export subsidy reduction being negotiated.

The proposal was sent to the CoAM for formal endorsement as a negotiating mandate. The next month contained seven CoAM meetings (only four of them serious) and one European Council meeting on the subject of the proposal, and represented the 'most emotional and protracted farm policy debates for years' (FT, 27 October 1990). The CoAM first considered it on 8 October. Only the UK, the Netherlands and Denmark initially agreed to it as a mandate for negotiations. Crucially, the French and German governments rejected it, the main criticism being that no study had been done by the Commission of the impact of the 30% cuts, nor any proposals suggested as to how any adverse effect on farm incomes could be relieved.

For a month, the agenda of the Trade and Agriculture Councils consisted solely of the need to agree a negotiating mandate. There was a classic game of 'buck-passing' between the two councils. The CoAM passed the issue to the Trade Council after the 8 October meeting. The Trade Ministers refused to consider the proposal without an official opinion from the CoAM. The FT (11 October 1990) noted that 'paralysis has gripped the EU'.

The German position hardened in October as the German national elections (in December) approached. In the 15 October Council meeting, the German farm minister, Keichle, opposed the 30% cut unless compensation measures had been agreed in advance. This line had been agreed by Helmut Kohl. *AgraEurope* (no. 1411) said of Keichle that he was 'not going to sacrifice his farmers on the altar of free trade'.

An agreement was almost reached at a CoAM meeting on 26 October. The German demand was met by a clause that the Commission would undertake 'to submit, rapidly, concrete proposals' which would reorientate CAP spending to support the incomes of farmers at a 'viable' level. After 16 h the meeting collapsed after Louis Memaz (France) balked at the effect on Community preference and import penetration in the EU. The EU could make 30% cuts in support levels by production control rather than price cuts; when tariffs were cut this would give imports to the EU a proportionately greater price advantage. Memaz proposed that reductions in tariffs be linked to reduction in domestic prices.

The Rome European Council session of 27 and 28 October collapsed over failure to get the agriculture part of the EU's 'final' negotiating position in the Uruguay Round agreed (it went on for another 3 years). Delors delivered a rebuke to Margaret Thatcher for her exhortation to get the agriculture issue sorted out: 'we are not the Americans, who negotiate and then consult their Congress'.

Agreement was reached at the 6 November CoAM on the basis that reductions in tariffs could be no quicker than the reduction in domestic support prices. The proposals for a 30% reduction in support, tariffs and the value of export subsidies were maintained.

Against this EU turmoil, the USA was able to achieve another propaganda victory by presenting its final package to the GATT secretariat on 12 October before the deadline set at Houston. This proposal was submitted under the three headings the USA had been using throughout the round. To look first at internal supports, red-box subsidies should be reduced by 75% from a 1986–1988 average over 10 years from 1991/92. Amber-box subsidies would be reduced by 30% with the same conditions as for red-box ones. All these reductions would be commodity specific, but it was allowed that an AMS might be used to 'express and monitor' the overall effect. Under the border protection list, all NTBs would be turned into tariff equivalents (TEs), bound and reduced by 75% over 10 years. Import access would be improved by the use of tariff quotas using 1986–1988 imports as a base, and expanded by 75% over 10 years and then removed. Further, the submitted proposal targeted export subsidies for reduction by 90% from a 1991–1992 average in terms of both aggregate budget outlay and the total quantity of exports assisted.

The chasm that existed when the EU finally made an offer was great. The USA would not accept the EU's unwillingness to make quantitative commitments on border protection and export subsidies. There was no acceptance of the EU's argument that agreement to reduce internal subsidies alone would also mean commensurate reductions in border protection and export subsidies.

In the period up to the final Brussels meeting the EU was very much on the defensive. The FT (28 November 1990) stated that the EU appeared 'forced to defend a position that seems timid and protectionist and is highly unpopular with all save a small but vociferous interest group'.

4.4.6 Brussels, 4–9 December 1990

Predictably, from these positions, no decision was reached. MacSharry had no flexibility in agriculture and the USA was not willing to compromise on the principle of the three-headings approach and quantitative limitations. A last-minute paper by Hellstrom, the Swedish Agriculture Minister, was a microcosm of the whole Uruguay Round, in terms of agriculture, up to that date. The proposal was a reduction of 30% over 5 years from 1991 based on 1990 support levels. Also, there would be a 30% reduction in export subsidies over the same time period using the 1988–1990 average. Hills accepted the offer as a 'basis for negotiation'. It was rejected by the EU; the MacSharry offer was at the extreme of the 'politically feasible'. This part of the Uruguay Round is important because of its timing: the MacSharry reform proposals were introduced within the Commission the week after the collapse of the Uruguay Round at Brussels. This period is covered in greater detail in Chapter 6.

4.4.7 Introduction: Uruguay Round 1990–1992: From Brussels to MacSharry

The history of the Uruguay Round after the original deadline had been missed at Brussels is centred on the proposal of Dunkel's Draft Final Act on 20 December 1991. This section highlights that proposal as the basis of the resolution of the outstanding issues. Any variable considered significant in the completion of the agriculture part of the Uruguay Round must be ranked according to how it relates to this Final Act. Chapter 7 completes the history of the Uruguay Round from 1992 to 1993, describing how the enactment of the MacSharry reforms by the EU affected the subsequent negotiations.

4.4.8 Background to the Draft Final Act: the Restart of February 1991

Dunkel announced the resumption of the Uruguay Round on 19 February 1991. The participants had agreed to negotiate 'specific binding commitments' under the three headings used hitherto only by the USA: import access, internal support and export subsidy. This appeared to be a major concession by the EU. The view by Dullforce (FT, 22 February 1991) was that this represented, for the first time in the Uruguay Round, a well-defined objective for negotiations. However, the extent of the concession can be judged by the fact that the EU did not publicly change its negotiating position through 1991.

On 4 February 1991 an initiative on CAP reform was promulgated by Agriculture Commissioner MacSharry (COM (91) 100). The formal progress of the MacSharry reforms is the subject of Chapter 6. These are described by Ingersent, Rayner and Hine (1993) as 'in the spirit of MacSharry'. However, the important point about this juncture is that a dialogue within the Community had been started; its apparently symbiotic relationship with the Uruguay Round is the reason for this whole section on the round as a causal factor. *AgraEurope*'s Green Europe supplement predicted as early as February 1991 that the EU's internal reform plans would be 'one of the most potent cards in the whole poker game' of Uruguay Round agriculture negotiations.

4.4.9 February–December 1991

The period from February to December 1991 was characterized by slow progress on the agriculture section of the Uruguay Round. The Commission's policy-making focus was on CAP reform and the 1991/92 price package. This fact started a debate about the ordering of GATT negotiations and CAP reform. Peter Lilley, the UK Secretary of State for Trade and Industry, insisted that a successful conclusion of the round must precede CAP reform; if the sequence were the other way round the Uruguay Round would be threatened by the EU's 'limited capacity for flexibility' (FT, 3 August 1991). This 'sequencing' debate is a crucial one in trying to define a relationship between the Uruguay Round and MacSharry reforms.

A certain optimism was generated by a meeting between George Bush, Jacques Delors and Ruud Lubbers (President of the Council of Ministers) in The Hague on the 9 November. This was the first high-level political intervention since the collapse of the Uruguay Round at Brussels in December 1990. EU officials were quoted in the FT (20 November 1991) as advertising the possibility of a deal on agricultural trade. USA officials were more cautious but agreed that the Bush intervention had relieved a 'political logjam'. The talk was of 30–35% cuts in the three areas over 5 years. Disagreement centred on the base period for the measurement of the cuts (the USA wanted 1986–98 against the EU's demand of 1986–90), the implementation period, whether export subsidy cuts would be measured by budget allocation or export tonnage, whether the DIPs envisaged in the MacSharry Plan of July 1991 (COM (91) 285) would be placed in the 'green box' of exempt subsidies, and finally, the EU's demand for rebalancing.

The optimism was dissipated as soon as the technicians went to work on the areas covered in the Hague meeting. Legras (head of DG VI) left Geneva after 1 day of meetings with Richard Crowther (USA Agriculture Under-Secretary). Disagreement festered on the quantitative limitation on subsidized agricultural exports. The USA proposed that two-thirds of the 35% cut in export subsidy support should be in the form of the volume of subsidized exports. The EU would be constrained to export a maximum of 11 Mt of cereals by the end of the implementation period. This compared with 20 Mt in 1990, 23 Mt then estimated in 1991 and a 1986–90 average of 17 Mt. The USA made the point that subsidized cereal exports had stood at 13–14 Mt when the round began in 1986 (all figures from the FT of 6 December 1991).

4.4.10 Dunkel's Draft Final Act, 20 December 1991

The Draft Final Act was the key proposal in the development of the negotiations on agriculture in the Uruguay Round after 1990. It effectively formed the base for the final agreement. In terms of assessing it as a causal factor in the passage of the MacSharry reforms, it is the reason why the beginning of 1992 differed from the beginning of 1991.

This proposal had a common implementation period (1993–1999) but two different base periods. Internal supports and border protection reductions would be measured against the average for 1986–1988, but export subsidies would be measured against the 1986–1990 average, thus splitting the difference of the demands described above. The Final Act was organized under the three headings: market access, internal support and export subsidies.

Market access was subject to an average 36% reduction in customs duties including those NTBs which had been tariffied. There would be a minimum 15% reduction in any line. Further, there would be minimum access opportunities, initially set at 3% of the importer's base period consumption and rising to 5% in the final year of implementation.

Under internal support reductions contracting parties were required to implement a uniform 20% cut in an AMS measurement of subsidies. The AMS defined was

$$(P_s - P_r)Q_s$$

where P_s = internal price, P_r = external reference price defined in domestic currency and Q_s = production.

This measurement did not 'credit' the EU for supply control to the level that its SMU would have done. Here P_s is replaced by a shadow price which reflects the imposition of a quota or some form of production limit (for example, the stabilizer regime's MGQ). Hence, measured AMS would have fallen in the EU before the internal price (P_s in the Draft Final Act measurement) was adjusted at all.

This section of the Draft Final Act also included a set of criteria for DIPs to qualify as decoupled from production, and therefore green box. If DIPs were to be related to prices, production or the employment of factor inputs, then only a fixed base period could be used. Also, eligibility for payments could not be made conditional on continued production after the base period. The export subsidy heading called for a 36% reduction in budget outlays on export subsidies and a 24% reduction in the volume of subsidized exports.

The main differences between the Draft Final Act and the EU's official negotiating offer (Brussels, December 1990) are as follows. The base period was changed from 1986; the EU had informally been moving away from this point through 1991. The method of measurement of that baseline support level was changed. The Draft Final Act outlined 'specific binding commitments' in three areas, not just in domestic support reduction. Further, there was an asymmetry in the actual reductions in these three areas: the market access and export subsidy constraints were larger than that for internal supports. The 20% cut demanded here seemed far from binding compared with the EU's offer of 30% in December 1990. Such a conclusion is supported by the fact that it was the proposals under the export subsidy heading which were the substance of negotiations on the Draft Final Act. This comparison of these two positions 1 year apart is instructive, because it defines the step the EU had to make to achieve agreement in the Uruguay Round. An assessment of the influence of the MacSharry reforms on making that step can be made from this.

The USA agreed to the Draft Final Act as the point from which negotiations should proceed. The EU held back from agreement with this; the 23 December CoAM meeting called the Act 'not acceptable', and said that it had to be modified. It was unacceptable on a number of counts. First, the market access proposal did not include rebalancing, and tariffication did not include a margin of Community preference. Second, as described above, the AMS measurement did not give the EU's demand on credit for supply control. Third, and the point which would rumble through the rest of the Uruguay Round, the EU did not like the volume requirement in the export subsidy reduction programme. The final point was that the Dunkel requirements for green-box status would exclude the DIPs envisaged in the MacSharry proposals for CAP reform.

The CoAM meeting of 10–11 January was considering CAP reform and GATT side by side. This was against a backdrop of protestations by senior commissioners and Council members that the two issues were not linked. The UK, German, Dutch and Danish farm ministers were very keen for GATT agreement. Brian Gardner in *AgraEurope* (24 January 1992) speculated that after the Council meeting of 24 January MacSharry was considering asking the College to change CAP proposals to fit in with the GATT negotiations. Specifically, there was a discussion about extending the transition period of COM (91) 258 from 3 to 5 years and a consideration of time-limited compensatory payments that would qualify for the GATT's green box, i.e. be judged as production neutral. *AgraEurope* (24 January 1992, p. 2) commented, 'It is quite clear that the Commission's thinking, in trying to make the reform plan more "GATT-able", is moving along these lines.' This is something MacSharry vehemently denied in the interview with him for this work in October 1994.

The Council's deliberations on 27 and 28 January 1991 were described by *Agence Europe* and *AgraEurope* as muted. This reflected Ministers' desire not to talk about CAP reform and GATT together. *AgraEurope* (31 January 1992) reported that the majority view at this meeting was to the need to reach agreement on CAP before making any concessions in the Uruguay Round. The Council's basic position of 23 December on the Draft Final Act remained and effort was employed in the first 6 months of 1992 towards the agreement of the MacSharry reforms.

4.4.11 The MacSharry Reforms, 22 May 1992

The Council of Ministers enacted CAP reform based on the proposals of July 1991. In some respects the MacSharry reforms exceeded the requirements of the Draft Final Act. Subsidized EU cereal prices would be reduced by 29% over 3 years. This reduction in turn would allow subsidized beef prices to fall by 15%, pork to fall by 15% and butter to decline by 5%. These reductions in internal support were expected to eliminate the need for most export surpluses by 1996–1997, hence the argument offered by MacSharry at the time of the enactment of the reforms that export subsidies would be eliminated rather than reduced as under the Draft Final Act, and that this would be achieved by 1996–1997 rather than over the 6 years of the Dunkel plan. Export subsidies were 'the high octane fuel of the GATT row' according to Gardner (FT, 22 May 1992). However, the idea that CAP reform addressed this issue was a gloss added by the Commission.

The EU was unwilling to agree to the 24% volume cut in subsidized exports proposed in the Draft Final Act. The MacSharry reforms, by providing for compensatory DIPs to farmers, opened up another area under the Act: the question as to whether these payments would qualify for green-box exemption.

This section has covered the history of the Uruguay Round in terms of being a cause of and influence on the MacSharry reform process. Only that part of the round which happened before May 1992 is therefore relevant to and included in this chapter. Chapter 7 contains the story of the Uruguay Round after the MacSharry reforms had been enacted.

4.5 CONCLUSIONS

This chapter breaks down the causes of the MacSharry reforms into three groups. First, there are the causes which are internal to the EU, namely the operation of the CAP in the period before the MacSharry reform process started. Second, the oilseeds dispute was a pressure for CAP reform from outside the EU. The third group of causes, also from outside the EU, are those associated with the negotiations of the agriculture section of the Uruguay Round. This chapter has outlined each of these groups of causes in terms of which part of the EU decision-making system they affected, along with how and when they affected that part of the system, so as to provide a description of the environment within which the MacSharry reform process took place.

There are two significant differences between the pressures for CAP reform emerging in the late 1980s and early 1990s, and the pressures which led to the milk quota and stabilizer reforms. First, there was the operation of the stabilizer regime, which had not had the desired effect of controlling EU cereals production and the related budget consequences, but which was yet imposing politically difficult automatic price cuts. Second, there was external pressure for reform. The EU was involved in international trade negotiations the agenda of which was dominated by the USA's demand for substantial reductions in the level of border protection, internal support and export subsidies.

NOTE

1. The Export Enhancement Program was a system which granted a proportion of public stocks free to traders for export to markets where there was competition from the EU.

Chapter 5

The Public Choice Paradigm of Decision-making Systems

5.1 INTRODUCTION

The increasing number and scale of interventions by governments in their economies since the Second World War has inevitably generated a commensurate research interest. There is a vast, varied and long-standing field of scholarship pursuing this interest. The pertinent research material here is the answers to the questions of why economic policies are changed, when they are and the way that they are.

This chapter describes the public choice paradigm alternative to the neo-classical economics approach to the issue of CAP reform. Section 5.1 develops the public choice approach, while Section 5.2 discusses the diversity of models within the public choice paradigm and Section 5.3 looks at the application of public choice research to the question of explaining the reform of the CAP. Section 5.4 provides a justification of the public choice paradigm as appropriate to use to generate common analytical frameworks to employ in the understanding of the MacSharry reforms.

Three frameworks for understanding the CAP reform process are discussed in Sections 5.5–5.8. Section 5.9 selects the institutions framework to confront with the evidence of the MacSharry reforms case study in Part II of this book.

5.1.1 Introduction to the Public Choice Paradigm

The public choice nomenclature is equivalent to the *new* political economy, or the economic approach to human behaviour, to borrow Becker's phrase (Becker, 1976a). It is also in certain literatures the equivalent to rational choice theory and social choice theory. Mueller (1989) defines a public choice approach as the 'economic study of non-market decision making, or simply the application of economics to political science'. Economics applied to politics is the theme adopted by Becker (1976a) and

Northglass (1986). Becker (1976a) borrows the famous Robbins definition of economics, 'the scientific study of the allocation of scarce resources which have alternative uses', to suggest that the notion of scarcity is equally applicable to the non-market sector of an economy. The analogy of the market from micro-economics can be extended to a political market for government intervention and regulation.

The public choice paradigm has three basic premises. These are the assumptions which cannot be challenged or altered without changing the paradigm. The first is methodological individualism. The unit of explanation of observed social phenomena is the individual. Public choices – that is, choices by any component of the political system – are ultimately made by individuals, and this is the basis of any explanation of a decision that has been made.

The second premise of the public choice paradigm is that individuals are rational actors. The three strict requirements of a rational actor are described by Elster (1986). The agent knows of a set of feasible alternatives. The agent has a set of beliefs about the causal connection of the agent's choice to an outcome. The agent ranks subjectively these outcomes and chooses an action from the feasible set which is expected to lead to the highest-ranked outcome. The second requirement of the rational actor is that preferences are stable, so as to avoid the conundrum that most actions can be justified *ex post* as rational in terms of 'this is what the agent actually wanted'. Third, opportunity cost is the standard by which ends are chosen, and the non-market sector may often use shadow or imputed prices.

The third premise of the public choice paradigm is that the preferences of individual agents in a political system are to some extent dictated by their position in that political system. This allows the modelling technique of the representative individual to be used. An abstract representative individual can be nominated for some component of the political system, an interest group for example. This individual's preferences can be stated as being conditioned by being a member of that interest group without any detailed knowledge being required about that individual. This relieves the research burden of accounting for how an individual agent in the political system has formed and developed his or her actual preferences. By using this technique the key mechanisms of interest group membership or actions can be studied.

This third premise provides for a great diversity of public choice models. There are a large number of different conceptions about how political systems function. The subject of comparative politics examines how political systems differ, and on what basis they can be compared. Further, even for a given political system there can be strongly divergent views on where power lies and the relationship between different components of the system. This point about the diversity of public choice models is considered in Section 5.2.

The public choice paradigm sets itself against naive empiricism, described by Bernstein (1976) as the lack of an explanatory foundation when correlations break down. Instead, the public choice paradigm seeks to support models which extend 'the scope and methods of debate and research in political science towards new forms of logically and mathematically informed reasoning' (Dunleavy, 1991, p. 259). The public choice approach also casts itself as the formalizer and organizer of arguments about observed social realities. It is a paradigm of how to analyse a decision-making system. Within this paradigm a number of frameworks can be developed. As is emphasized throughout this chapter, public choice analysis provides no definitive answer, just a paradigm within which frameworks can be developed and considered.

5.1.2 Two Strands of Public Choice Research

Dunleavy (1991) and Green and Shapiro (1994) note that a bifurcation has occurred in the field of public choice between the elaboration of theoretical rational actor models and empirical applications of public choice. There is a large field of complex theoretical models in the public choice paradigm, which are based on restrictive assumptions about the degree of rationality and how political systems work, and therefore seem to have little relevance to the 'messy' empirical world of government decision-making.

The second field of public choice scholarship attempts to apply a public choice approach at an empirical level: to explain observed public choices. This class of public choice models aims to 'offer a compelling, applied and relatively detailed account of how the core processes of liberal democratic politics operate' (Dunleavy, 1991, p. 2). They eschew the abstraction of the first strand of the public choice paradigm; to repeat, they use a public choice approach to try to explain observed government decisions. The MacSharry reforms are an observed outcome of the EU agricultural decision-making system. Thus this work is located in this second strand of public choice work.

The split between the two strands of public choice scholarship is not so complete that there is no cross-over. Insights and concepts developed in the abstract field have been used to try to gain an understanding of government decisions. Therefore, the literature of the theoretical, abstract part of the public choice paradigm is briefly surveyed in Sections 5.2.1–5.2.3.

5.2 THE DIVERSITY OF MODELS AND FRAMEWORKS WITHIN THE PUBLIC CHOICE PARADIGM

The third premise of the public choice paradigm, that individual preferences are governed to a large extent by that individual's position in the political system (Allinson, 1971), allows different specifications as to how the political system works to be introduced. Different conceptions of the important elements and relationships in a political system can be used within the public choice paradigm.

This point leads into Sections 5.5–5.9, which discuss three different frameworks for understanding the CAP decision-making system. These are based on different understandings of that system. One of the objectives of this work is to compare these different frameworks and choose one as superior in analysing CAP reforms. All three frameworks are within the public choice paradigm and a discussion of their relative merits does not imply an attack on the paradigm.

5.2.1 The Diversity of Public Choice Models: Examples from the First Branch of the Discipline

A brief examination of some of the central works of the first strand of scholarship in the public choice paradigm suggests a divergence in the theories of the state[1] employed in models. The different frameworks for understanding the functioning of the state can be divided into those that view the state as an independent agent or series of independent agents, and those in which the state's activity – that is, public policy – is seen as the outcome or result of the interaction of the political system. The former category can be further divided into those in which the state acts as an omniscient, benevolent, social welfare maximizer, and those of the *state-centric* approach in which the state pursues its own agenda, separate from any social welfare function, within constraints imposed by the society. This basic dichotomy is suggested by Przeworski (1985). In the political economy literature, a similar distinction is made by Bhagwati (1982) between models with a *self-willed* government maximizing some objective function and a *clearing-house* government which responds to the pressures from the political system.

The state as social welfare maximizer is simply a translation of the usual rational economic agent with a given preference function to government level with society's aggregate preference function. The usual Arrow's Impossibility Theorem dilemma is usually circumvented in the models by allowing dictatorship. Riker (1982) gives a clear outline of Arrow's proof. (Arrow's Impossibility Theorem is a proof by Kenneth Arrow that it is impossible to devise a constitution or voting system, complying with certain reasonable conditions, which can guarantee to produce a consistent set of preferences for a group from the preferences of the individuals making up the group. The system which *can* produce a consistent set of preferences is to allow one individual to always make the social choice – a dictatorship.) The benevolent dictator model describes the state as choosing public policy, in the same way that a rational economic agent might spend the family budget.

The state-centric approach, in its most complete form, has the institutions of the state having solved the free-rider problem of collective action

(Olson, 1985) both within the institutions of state and between the institutions of state. The state is still a unitary decision-maker, and an agenda is pursued and maximized subject to the constraints imposed by the ability of other organizations to muster political and economic power against the state. A less demanding interpretation has the state as a series of institutions that are competing with each other to control aspects of public policy, and that will use alliances with other groups in society to push issues and causes against other institutions of the state. This view has large scope for applying the techniques of game theory; state actions can be understood as the outcome of a series of games between institutions of the state and between those institutions and groups in society. The distinguishing feature of this version of the state is that the state controls the agenda of public policy and initiates public policy. It manages to do this by controlling information, monopolizing the use of force and centralizing economic power – the usual litany which comes under the title of state power.

This concept of state contrasts with the view of the state as an abstract from the society below and whose power is *captured* by competing groups. Interests are said to compete for the rents that can be created by state intervention. The general point for this section is that public policy is the outcome of the political battle of various interests in society and the state itself is not an agent but rather a source of power to be won or controlled.

5.2.2 State-centric Approach

There is a class of public choice models in which there is a self-willed government maximizing some weighted social welfare function. This is sometimes called a political or policy preference function (PPF). Rausser and Freebairn (1974) were the first to provide empirical estimates of a PPF (in their case the self-willed government of the US beef industry). They imputed the welfare weights in the PPF given to each sector affected by US beef import quotas. In an EU context, Burton (1985) used a PPF approach in the dairy sector to predict the level of milk quotas. Despite these empirical applications, the PPF approach belongs to the first strand of public choice literature. The underlying political process which produce these weights in the PPF is not accounted for.

5.2.3 Rent-seeking and Clearing-house Models of Government

Rents are returns on factors of production in excess of that level of return required to keep that factor employed in its present position. Governments can produce rents by intervening in the economy. Interventions of this kind are thus desirable to those owning such factors. Hence, there exists a market in political support; interest groups or individual voters provide different forms of political support, from votes straight through to bribes, to politicians who respond by providing interventions and associated rents. Examples abound; Krueger (1974) describes the competition for import licences in India as one of the clearest examples of rent-seeking.

This literature is divided on the relative importance attached to interest groups or individual voters in producing the political pressure for politicians to respond with interventions in the economy which produce rents. Brooks (1995, 1996) and de Gorter and Swinnen (1994, 1995) disagree on the issue of whether voters are rationally ignorant of the effects and the costs of agricultural policies. The rationally ignorant voter was described by Downs (1957). Government interventions may have concentrated benefits and spread costs widely. The more diffuse the incidence of the costs of the intervention, the less incentive there is for the individual voter to learn about that issue. The voter remains rationally ignorant.

Brooks (1996, p. 370) states that the assumption that voters are rationally ignorant is 'intuitively central to an understanding of why most OECD countries subsidise their farmers'. The application of economics to politics has consistently produced the conclusion that the information flow in the democratic process is filtered and noisy when compared with the economic market (see Riker, 1982). Voting is infrequent, one vote encompasses a series of issues, and everyone votes, not just those directly associated with the decision. In the economic market agents do not vote on matters of no particular interest to them. Further, information is cheap (classic supermarket analogies apply here) in an economic market. This kind of market is a superior mechanism for revealing sovereign preferences, because people feel the full cost and benefit of their decision (assuming away externalities). In a world of rationally ignorant voters, interest groups are the key actors in the political system. Through collective action they produce political support and receive rent-creating interventions in return from politicians (otherwise known as pork barrel politics). In his model of interest groups, Becker (1983, p. 372) notes that 'voter preferences are not a crucial *independent* [my italics] force in political behaviour'. Rather, it is the observation that farm interest groups are relatively more efficient (compared with other interest groups) at producing political pressure that explains the structure of agricultural policies in OECD countries.

De Gorter (1994) and de Gorter and Swinnen (1994, 1995) argue that the activities of interest groups are not a fundamental factor in determining the structure of agricultural policies across OECD countries. Voters are rational, self-interested and fully informed about agricultural policies, rather than rationally ignorant. An understanding of the nature of agricultural policies in the OECD requires an analysis of the interaction of political support-seeking politicians and support-supplying voters. The politician–voter link is more important than the politician–interest group link.

5.3 PUBLIC CHOICE APPROACHES TO CAP REFORM

The literature looking at the specifics of a public choice approach to the CAP decision-making system belongs largely to the second strand of public choice paradigm research. Olson (1965, 1977) and Becker (1983) assert that farmers in industrial countries are relatively more efficient at producing political pressure than consumer groups, and therefore agricultural policies tend to favour farmers. This is a claim taken up by Senior-Nello (1984) and Averyt (1977) with regard to the CAP. The farm lobby has historically been strong in the member states and at a Commission level. This is because European farm lobbies are well organized and well resourced and have solved the free-rider problem (Olson, 1985). The concepts of *exit* and *voice*, introduced by Hirschman (1970), have been used here; the economic structure of agriculture means that farmers have limited exit from the industry, therefore as a group they tend to rely on voice – that is, political activity – to sustain their livelihood.

Koester (1978) and Schmitt (1986) focus on the institutional peculiarities of the CAP decision-making process to explain a bias towards the farm interest and against consumer interests. This fits in with work done in the first strand of public choice on the importance of voting rules and the organization of committees (Riker, 1982; Mueller, 1989). Senior-Nello (1984) and Schmitt (1986) highlight two peculiarities in the CAP decision-making system which affect the type of agricultural policy decisions taken and the manner in which they are made. The first is the operation of the convention of unanimity after the Luxembourg compromise of 1966. The easiest way to reach agreement in the annual price review among countries with different agricultural sectors and different inflation rates is to keep raising the common price to the extent that each member state is satisfied, at least to the extent that it will not employ the veto. This leads Schmitt (1986) to the conclusion that the increase in the level of prices under the CAP has been greater than would have been the case if there had been a series of national agricultural price support policies (the counterfactual of there being no CAP).

Petit *et al.* (1987) make the further point that the unanimity principle in the CoAM means that any one member can block an issue, so issues are forced into packages by a form of politicking known in USA politics as *log-rolling*. The argument runs that there is a *core* to any package of issues – those issues that have a high cost to some countries and on which those countries might credibly threaten an indefinite veto. Around the core are 'peripheral' issues which are important to a small number of countries and which may be traded for positions in the core. The concerns of national ministers of agriculture about how the national constituency will react to any decision means that Council meetings can frequently be drawn out in order to give the impression of last-ditch defending of certain interests, even when those interests have already been conceded to achieve agreement. Smith (1990), with regard to the US Congress, calls it the *blame game*.

The number of CAP decisions taken according to the need to achieve unanimity has diminished substantially since Schmitt and Senior-Nello wrote their articles. The CoAM negotiations which led to the enactment of the MacSharry reforms were conducted by the president and the Commission, on achieving a qualified majority. The reintroduction of qualified majority voting in CAP decision-making (provided for in Article 148 of the Treaty of Rome) came at the same time that the Single European Act of 1986 provided for its increased use for measures to complete the single market.

The second peculiarity of the mechanics of CAP decision-making highlighted by Senior-Nello (1984) and Schmitt (1986) is the ability of the CoAM to spend without reference to Ecofin or the effect on the total EU budget. As noted in Section 3.4, this was a prime motivation behind the Delors package of budget measures agreed in February 1988. The introduction of budget guidelines and a medium-term target for the growth of agricultural spending has altered the situation which had existed in the 1970s and most of the 1980s: the CoAM having the ability in its Annual Price Review in March/April to take decisions with financial consequences greater than any previously agreed budget limits. Schmitt (1986, p. 338) describes a Treaty base 'which states that "obligatory expenditures" have automatically to be financed by the budget. Financial expenditures connected with farm policy decisions are in almost every case such obligatory expenditures.'

5.4 RATIONALE FOR THE USE OF THE PUBLIC CHOICE PARADIGM IN THIS BOOK

Sections 5.1–5.3 set up the public choice paradigm as an alternative to mainstream agricultural economics in the explanation of CAP reforms. It is employed in this work for two main reasons. First, it can provide different frameworks for understanding CAP reforms. This flexibility of the public choice approach is advantageous when analysing CAP reforms which have occurred at different times over a long period. Second, the public choice approach suits the issue of CAP reform. As argued by Hagedorn (1985), CAP reform should be seen as the result of a process. Within this process there are a number of links that can be established. The three premisses of the public choice paradigm encourage this search for smaller and smaller links, or reduction in the time between causes and effects; the whole thrust of a public choice approach is to break down a collective decision into its constituent decisions made by individuals. To understand the CAP reform process requires a disaggregated perspective, and the public choice paradigm encourages the disaggregation of observed government choices.

5.5 THREE FRAMEWORKS FOR UNDERSTANDING THE CAP REFORM PROCESS

The three frameworks, all within the public choice paradigm, for understanding the CAP reform process are presented in this chapter: the interest groups, the prominent players and the institutions. Each of the frameworks shares the position that political systems can be discussed in terms of a state and a polity sectioned into groups. They differ in their conception of the state and the nature of the relationship between groups and that state. This is the same point made in Section 5.2 about the diversity of frameworks within the public choice paradigm.

The *interest groups* framework holds that the state is an abstract entity, a black box whose power is captured by groups competing in the polity. The socio-economic interests affected, or potentially affected, by state interventions compete for control of the state. The state is not itself a political actor, but a reflection of the competition of interest groups. Any analysis of what legislative decisions have been made, or will be made, must start from considerations of group competition. Moyer and Josling (1990) call these 'outside political inputs'.

The *prominent players* framework and the *institutions* framework both admit the state as a political actor. Further, both aim to open the black box of the state; the state is not a unitary actor, rather it is a set of institutions competing for legislative, bureaucratic and financial control of public policies. In common with the interest group framework, both assume that there are organized interests in society that are affected by state intervention and hence are interested in affecting state action. The prominent players and institutions frameworks both start from the position of a fragmented state and a plurality of interests competing for state action. They differ on how they define that relationship.

The prominent players framework is the view that CAP legislation should be viewed in terms of the outcome of the interaction of prominent players. In any public policy sphere, certain interest groups and certain state institutions are listed as prominent players. Their web of interrelationships should be examined to see how institutional structures affect interest group influences on CAP decision-making.

Prominent interest groups enjoy the political power to force decisions to be made or not made. However, an integral part of their influence is the privileged institutional access they enjoy. Their relationships with the institutions of the state are central to understanding agricultural policy. Institutions matter because they affect 'the structure, scope and character of activity by interest organisations' (Grant, 1993, p. 44).

Smith (1993) provides a typology of these state institution–interest group relationships, running from closed, stable policy communities (Richardson and Jordan, 1979; Richardson, 1982) which have been applied to the EU generally and agriculture in particular (Smith, 1990; Grant, 1993) to open, unstable issue networks. The histories of the CAP reforms of 1984 (Petit *et al.*, 1987) and 1988 (Tracy, 1989; Moyer and Josling, 1990) have been written within the prominent players framework.

The *institutions* framework holds that when, how and what legislative decisions are made about the CAP are functions of the configuration of the path along which any legislative proposal must pass. The path begins where the power of proposal exists and finishes where the power of veto resides, or the power of formal enactment. It is the configuration of all those who hold some claim on the policy process; those components of the political system that are in a position to claim some stake in, and influence over, the proposed legislation.

The institutions framework holds that the actors with the greatest influence in the black box of CAP decision-making are the institutions directly involved in CAP policy. In contrast, interest groups are able to achieve only a very limited influence in the legislative process. They are not central to understanding why CAP reforms occur when they do, and in the way that they do. CAP reforms are conceived, constructed and enacted within the black box of state institutions. The institutions involved have their own agenda for, and compete for control of, public policies. This competition cannot be reduced to the pressure from interest groups, nor are interest groups important allies in this competition.

The remainder of this chapter sets out the different frameworks. The arguments for and against each framework's relevance to the understanding of CAP reforms are presented. This will serve to highlight the key differences between the three. Section 5.9 selects the institutions framework to confront with the evidence of the MacSharry reforms case study in Part II of the book, and justifies that selection.

5.6 THE INTEREST GROUPS FRAMEWORK

This section outlines the interest groups framework and traces and references its intellectual history. The criticisms of the framework are provided in Section 5.6.1, with attention paid to its failure to support explanations of actual CAP reforms.

This framework holds that the stability of the CAP is explained by the strength of farm interest groups in the EU. Their organization and political influence has allowed them consistently and successfully to capture the abstract state referred to earlier. The more general point is that it is group competition which drives public policy in the EU, including reforms of the CAP.

The use of the framework in agricultural policy analysis has two sources: (i) there is a literature of abstract, non-EU and non-agricultural policy work which has been applied to the CAP; (ii) the history of the development of state support for agriculture at an EU and member state level has been written in terms of interest groups.

The application to the agricultural policy of advanced industrial economies of general, abstract work completed on interest groups in political science and economics rests on the following calculation. There are well-organized interest groups in every advanced industrial nation, as well as state support for agriculture. Interest groups enjoy strong functional relationships with the state and governments. *Ipso facto*, there is a causal link from strong interest groups to well-funded, pro-farmer agricultural policy.

Agriculture is mentioned as an example of successful interest groups in the following works: Olson (1965) with the concept of free-riding; Stigler (1975) and Becker (1983) on the use of political markets; Hirschman's (1970) concepts of 'exit' and 'voice'. The farm lobby provides a ready example of a well-organized, disciplined group. Brooks (1996) provides a survey of the formal political economy models developed within the interest groups framework of agricultural policy analysis.

The second source feeding the dominance of the interest groups framework has been the histories of state support for European agriculture including the CAP. Tracy (1989) and Milward (1992) trace the construction of a common agricultural policy in terms of the political influence of national farm groups. The history of UK agricultural policy from 1947 to 1973 is told by Self and Storing (1962), Smith (1990) and Tracy (1989) as the history of the strength of the National Farmers' Union (NFU).

These histories also detail a system of institutional arrangements set up to provide a regular and routine interface between the farm lobby and government. Institutions are highlighted as an important factor in explaining the persistent influence of the farm lobby in the prominent players framework and the institutions framework (see Sections 5.7 and 5.8).

5.6.1 Criticisms of the Interest Groups Framework

The interest groups framework has been criticized on two grounds: its supposed inability to explain the stability of the CAP, and, more relevantly to this chapter, its unsatisfactory account of periodic turbulence and reform of the CAP.

The first set of criticisms have been based on empirical investigations of the distribution of benefits of the CAP. If farm interest groups determine the CAP, why have average farm incomes remained stagnant in real terms and the numbers of farmers in Europe declined? Howarth (1985) provides a detailed exposition of these arguments. Harvey (1982) uses the Newcastle model of the CAP and some rough proxies for national interest to demonstrate a negative correlation between a national interest in improving farm incomes and how much the CAP has improved farm incomes in each member state. (The Newcastle model is an economic model of the CAP developed at the University of Newcastle. Its output variable is aggregate farm income at a member state level. It has been used to estimate the national redistributive effects of various CAP policy scenarios.) The interest group framework asserts that the stability of the CAP is explained by the political strength of farm interest groups. The question, given these kind of results, is: why do interest groups allow the underlying stability of the CAP to persist?

These criticisms can be answered by arguing that farm groups are dominated by the interests of the leadership elite. A famous statistic quoted by Commissioner MacSharry during the reform process between 1990 and 1992 was that 80% of CAP payments were enjoyed by 20% of farmers. That 20% provides the leadership of national farm interest groups. This point in a UK context reaches back to Self and Storing's (1962) description of the leadership of the NFU as being dominated by representatives from large arable farms.

The second category of criticism is more basic. Work conducted using the interest groups framework has failed to provide an account of the reform of the CAP which has any intuitive appeal. This is true for a number of reasons. The literature has interest groups as the explanatory variable of CAP stability and therefore the difficulty mentioned in Chapter 1 occurs: using the terminology and concepts developed to account for stability to explain turbulence. An explanation of CAP reforms within the interest groups framework requires some account of a shift in the balance of power between interest groups associated with the CAP.

The frameworks presented in Sections 5.7 and 5.8 do not doubt the history of the CAP written in the interest groups framework. However, the argument is that the forces which initiate public policy are not necessarily the forces which sustain and operate an established policy. The interest groups framework has produced clear and compelling accounts of the history of the setting up of the CAP, but has yet to produce such accounts of the history of CAP reforms.

The application of the framework to CAP reform has generally taken place in the first strand of the public choice paradigm. This has led to the stylizing of the CAP decision-making system to achieve modelling coherence, which has led in turn to a reduction in the ability to explain observed facts about the periodic reforms of the CAP. This is the reason why the interest groups framework can sometimes appear to be an extreme form of pluralism, used by agricultural economists to construct abstract models of the black box of CAP policy decisions.

5.7 THE PROMINENT PLAYERS FRAMEWORK

The prominent players framework emphasizes the series of systematic contacts and relationships between farm groups and the state in illuminating this problem of turbulence and stability. This section details the prominent players framework and the degree to which it benefits an understanding of CAP reforms: Section 5.7.1 describes the history of the prominent players framework and its general view of how the CAP decision-making system works; Section 5.7.2 sets out the specific claims on which the application of the framework rests; Section 5.7.3 looks at arguments against those claims – this leads into the institutions paradigm in Section 5.8.

5.7.1 The Prominent Players Framework: an Overview

In the framework which suggests that the CAP policy-making system should be viewed in terms of prominent players, those players are the interest groups affected by the CAP and the institutions involved in the formal enactment of agricultural policy. From the point of view of explaining CAP reforms, they are prominent because they enjoy some control of the resources which are necessary to the enactment of a CAP reform, be they political, financial or technical. The basic assumption of the framework is that 'the major farm policies survive because of the particular sets of institutions involved in the setting of policy and the structure of the decision framework which they operate, as well as the pressure from interest groups' (Moyer and Josling, 1990, p. 45).

The main development from the interest groups framework is that the institutions set up to regulate and depoliticize the contacts between the state and farm interest groups at European and national levels are themselves prominent players. They affect the balance of the competition between agricultural and non-agricultural interests. The fact that a farm group's absolute political strength may have declined does not necessarily affect its influence, because it does not have to compete with other groups for government's attention and financial support. Schmitt (1986, p. 342) highlights the 'absence of effective checks restricting and repressing the unilateral influence and pressure of the agricultural sector'.

The framework uses the term *agricultural policy interest group* to cover a wider range of groups than implied by the term 'farm lobby'. Farmers, agribusinesspeople and retailers all have a large economic interest in the CAP. Further, these are supplemented by an increasing number of environmental groups placing pressure on CAP policy-makers. Cox *et al.* (1986, 1987), writing in a UK context, encourage this wider interpretation of the political pressure from the agricultural sector. They reject the view that agricultural policy outputs in the UK should be seen as the result of exclusive MAFF–NFU interactions (MAFF being the Ministry of Agriculture, Fisheries and Food). The prominent players framework makes no substantive claims about the relative influence of the interests within the agricultural policy interest group category; it claims simply that there is a complex set of different interests.

The system of state institutions administering support and agricultural policy interest groups can be described as a *policy network* (Marsh and Rhodes, 1992). Marsh and Rhodes provide a taxonomy of these policy networks. The type of policy network relevant to the prominent players framework is a policy community.

Richardson and Jordan (1979) and Mazey and Richardson (1993) describe a policy community as having the following characteristics: (i) limited numbers of institutions and groups are involved in decision-making; (ii) the membership of the community is small, and continuous across time; (iii) the members interact frequently, across all aspects of the policy and there is a high degree of consensus as to the means and ends of the policy; and (iv) the prominent players in a policy community regard power as a positive-sum game, so that for each member their influence is maximized by being a member of the community as opposed to remaining outside it.

Policy communities can be placed at one end of a horizontal spectrum of policy networks, the other end being *issue networks,* which possess none of the characteristics of a policy community defined above (Marsh and Rhodes, 1992). Smith (1993) suggests that it is also possible to separate policy communities vertically according to degrees of state autonomy. State autonomy is defined as the capacity of state actors and institutions to develop policies without the input of interest groups.

The term 'policy community' was originally developed to argue that groups had significant resources and could limit the extent of state autonomy; the existence of a policy community was an illustration of a lack of state autonomy. Richardson and Jordan (1979) provide a detailed account of this view. However, Smith (1993) reinterprets the existence of policy communities as evidence of the autonomy of state institutions. The agenda of, and the mediation of different views within, the policy community is dominated by state institutions. By establishing these communities with affected interest groups, state institutions actually increase their capacity to act autonomously.

5.7.2 The Specific Claims of the Prominent Players Framework

The prominent players framework has three components: (i) that the CAP decision-making system should be analysed as a policy community; (ii) that interest groups are prominent players and part of that policy community; and (iii) that the first two parts together will allow the questions of stability and turbulence raised in Chapter 1 to be satisfactorily answered.

Any attempt to describe the CAP decision-making system as a policy community starts from the point that it will be more extensive and complex than the relationship between a ministry and its client groups (i.e. farm groups, food processors, agribusiness interests) at the level of member state governments. The EU is a supranational level of government, hence interest groups affected by agricultural policy have two strategies: lobbying at the national level and lobbying at the European level. The first refers to the lobbying of member states' vote on the CoAM. The second, lobbying at the European level, refers to lobbying the Commission directly through membership of the umbrella farm organization, COPA. These two sets of relationships provide a wider policy community than those which exist in national policy arenas. The merits of following two levels of lobbying are discussed by Grant (1993) and Averyt (1977).

The role of COPA has been the subject of research. Averyt (1977) provides the general conclusion that COPA has a role only in the bureaucratic functioning of the extant CAP system. COPA usually has 50% of the membership of the Commission's advisory committees which exist for each product regime. Hence Smith (1990) states, 'COPA's relationship with DG VI is . . . very close', but limits this conclusion to issues in the administration of the CAP.

Petit *et al.* (1987), Tracy (1989), and Moyer and Josling (1990) do not give COPA any role at all in the reforms of the CAP in the 1980s. COPA was forced to react to the CAP reform agenda. In the reforms of 1984 and 1988 COPA was unable to agree a response when the proposals were promulgated. This failure affects any account of interest groups as prominent players in a CAP reform process. If COPA has strong links with certain parts of DG VI on certain issues, such as the sharing of information on how the CAP is functioning in different parts of the EU, this does not seem to extend to having a position in any policy community regarding the issue of CAP reform.

For agricultural policy interest groups to be prominent players and members of a CAP policy community, they must materially affect the reform process and the reform enacted. In other words, they must aid the explanation of why reforms occurred at that time and why the reforms were of that particular type. The studies of COPA listed above and the results of the fieldwork interviews conducted for this work suggest that the organization has had minimal influence over the timing or content of CAP reforms. If this level of lobbying is not successful for national groups, in the sense of making them prominent players, then the national CAP interest groups must affect national positions in the CoAM in order to sustain interest groups as part of the EU agricultural policy community.

Petit *et al.* (1987) list four factors which affect a national farm lobby's ability to manipulate its national government's positions: (i) the cohesion of the national farm lobby; (ii) the functional relationship with the national minister of agriculture; (iii) the importance attributed to agriculture by a national government; and (iv) the political power of the minister of agriculture within a national government. It is on the positive existence of these factors that the prominent players framework rests. When these conditions exist, interest groups are prominent players and can be said to exist in a policy community. Without these conditions it is difficult to claim that interest groups are prominent players. This is one criterion on which the applicability of the framework to the issue of CAP reform can be judged.

5.7.3 Arguments against the Prominent Players Framework

This section considers more general arguments which doubt the inference from (a) the observation that each national agriculture minister is committed to his or her nation's agriculture sector in a reform situation to (b) the contention that agricultural interest groups have influenced that commitment. The arguments can be marshalled around two questions: why do national agriculture ministers adopt pro-agricultural sector perspectives in the CoAM? and, given these perspectives, how are they exercised in a reform environment?

The institutions framework holds that a national ministry's view of the nation's agricultural interest will generally coincide with the interests of the economically strongest part of the agricultural sector. This position taken by national representatives in the CoAM is a result of the institutional fact that they are ministers of agriculture and that is their department's *raison d'être*. They have a predisposition towards a readily identifiable client group: the agricultural industry. Three of the four factors listed above (Petit *et al.*, 1987) are institutional. The relevant counterfactual question is: if decisions about the reform of the CAP were made outside the CoAM, would the farm interest be so prominent?

The four factors of Petit *et al.* (1987) provide a limited view of the relationship between a national farm group and a national agriculture ministry. Five questions can be set up, the answers to which will provide a more substantial account of this relationship: (i) can interest groups influence to any extent the agenda of the relevant national ministry? (ii) Does the ministry control the production and distribution of information? (iii) What is the distance between national farm groups and the institutions at an EU level? (iv) Who designs the language to describe the recent history of agricultural policy? (who controls the history?) (v) What resources do interest groups possess which are necessary for the enactment of a CAP reform? The issues raised by these questions were considered in all the fieldwork interviews for this work. The results emerge in Chapters 6 and 8 in the discussion of the institutions framework's interpretation of the MacSharry reforms.

The extent to which agricultural policy interest groups can influence the behaviour of the relevant national ministry in the CoAM can be limited. CAP reforms are, by their nature, compromises. These compromises are reached through political trading between each member state's initial bargaining position. The dynamics of this political trading are difficult for a national interest group to control. As noted earlier, successful national interest groups must have a dual lobbying role when dealing with European issues: through national representatives and direct to the institutions of the EU. A strategy concentrating exclusively on the national route runs the risk mentioned above: the manoeuvres of political trading run beyond their influence or control. This is the nature of the black box for European policy. National agricultural policy interest groups are at one remove from the centre of CAP policy-making and can influence neither the agenda nor the bargaining process. This is the claim which divides the prominent players and the institutions frameworks.

To preserve agricultural policy interest groups as prominent players, it is necessary to claim that these groups influence their national agricultural minister's position. The Smith (1993) interpretation of policy communities can support this claim. The CAP is a series of national policy communities. The important factors are the structure of national agricultural policy-making institutions and their attitudes to the aims and objectives of agricultural policy. These result from that member state's history of national agricultural support policies, experience of the CAP and the economic position of its agricultural sector. These institutions decide which agricultural policy interest groups to include in a policy community, and make certain elements prominent. These alliances between the group and the institution are based on the institution's agenda. The decision of an institution to include certain agricultural policy interest groups in a policy community reduces the constraints on the ability of that institution to act in the agricultural policy process (at national or EU level).

The counter-argument is that there is no single, national agricultural policy community. Jordan *et al.* (1994), in UK terms, talk of a flexible system of policy communities operating at a sub-sectoral level. Cox *et al.* (1986) talk of a 'beleaguered policy community' when considering the end of the NFU monopoly on political pressure and representation on agricultural policy matters. Further, there are a number of different national government institutions that are affected by the CAP; for example, the effects of the CAP have spilled over into the international trade and budget fields of a national government's interest. This complex system of policy communities exists in a system of conflict, with competing ambitions for, and views of, the CAP. The policy communities do not share beliefs, perspectives and information on CAP reform. There is no single UK agricultural policy community. Indeed, the complex pattern of state institution–agricultural policy interest group interactions is such that it can be doubted whether the term 'policy community' lends much to the interpretation of UK agricultural policy-making. The more relevant consideration is the degree to which agricultural policy interest groups enjoy the political resources to influence the output of agricultural policy. This determines whether agricultural policy interest groups are prominent players in the formation of national positions with respect to CAP reform. An important point, and one mentioned previously with respect to COPA, is that CAP reform is a one-off issue, so there is no established policy community. Hence, even if it is concluded that national agricultural policy interest groups enjoy influence in a policy community, the question remains whether the decision-making process surrounding CAP reform proposals exhibits those characteristics.

5.8 THE INSTITUTIONS FRAMEWORK

The central theme of the institutions framework is that the institutional context of the pressures for CAP reform is most important in understanding such a reform. The institutions of CAP decision-making are the transmission mechanisms from the pressures to reform the CAP (external or internal) to the enactment of certain CAP reforms at certain times.

This section provides an outline of that institutional context. This leads to a consideration of the claim that the institutions of CAP decision-making define the set of feasible policy options, and proscribe certain options for the reform of the CAP. The application of the framework can produce the conclusion that without a change in that context, radical reform addressed to the persistent economic problems of the CAP will not be enacted. This explains the property of stability discussed in Section 1.5.

5.8.1 Institutions Framework: an Overview

The framework highlights the path along which any CAP legislation must pass as the relevant area to research in order to understand the process of reforming the CAP. The following factors are material in the CAP legislation which is enacted. The first is a characterization of each of the individual institutions in the chain: the various interests within them, their internal organization, and the power they hold over the legislation which passes through them along the path. The second factor is the interrelationships between each institution in the path: the balance between constitutional powers and accepted conventions. The third material factor is the institution which arbitrates in the event of dispute between institutions in the path.

The path is a combination of the formal, legalistic and constitutional criteria, plus generally accepted government routines. Hence, *legitimacy* is often used in such analyses (Hagedorn, 1985). This is what Moyer and Josling (1990) call in their schema 'inside political inputs'.

The difference from the prominent players framework is that agricultural policy interest groups are not in the path and hence are not a significant factor in explaining CAP decisions. Peterson (1995) charts the rise of *new institutionalism*: 'EU institutions may develop their own agendas and act autonomously of allied interest groups' (p. 81). Interest groups will have access to these institutions, but this does not equate to genuine influence in a reform situation. EU agricultural policy can be enacted legitimately without the support of interest groups. This is not to argue that state institutions enacting agricultural policy do not wish to have the support of agricultural policy interest groups; rather the claim is that agricultural policy interest groups do not hold the resources necessary to influence or veto the CAP reform process.

This view of the autonomy of institutions differs from Smith (1993) (in the prominent players framework) in the degree of political resources attributed to agricultural policy interest groups. The Smith (1993) view is that these groups enjoy a level of political resources, influence or power such that institutions wish to include them in the policy process through a policy community. The institutions framework attributes much less influence to agricultural policy interest groups. Jordan *et al.* (1994, p. 519), in a discussion of the fragmentation of the agricultural policy agenda in the UK, admit that 'it could be argued that the continued privileged position of British farmers within domestic policy-making arrangements has . . . much to do with the privileged position farming enjoys in the European arena'.

The central institution of CAP decision-making is the CoAM. Twelve (now 15) agriculture ministers formally enact any CAP reform, hence all factors and mechanisms involved in the CAP reform process must be traced back from that point. To understand the CoAM it is necessary first to examine the rules of the CoAM; for example, what determines the agenda, who makes decisions, and what are the voting rules. Second, each member of the CoAM must be put into his or her political context (at both domestic and EU level). This will afford an understanding of their bargaining positions and 'domain of feasible compromise' (Petit *et al.*, 1987).

Allinson (1971) illustrates how the bargaining rules can affect the outcome of that bargain. The convention of unanimity has operated at various times in the CoAM. Runge and Von Witzke (1987) apply the *n-person veto model* of Mueller to the CoAM. This shows that the convention of unanimity should have led to an equalization of benefits from the CoAM at member state level. It is sufficient at this stage to state that the CAP reforms of 1984 and 1988 were agreed using qualified majority voting.

Formally, the Commission controls the agenda of the CoAM. It enjoys the sole right to propose legislation, and for the CoAM to agree on something different from the initial Commission proposal requires unanimity. Hence a reform situation is created by the Commission by the proposal of reform. Peterson (1995) calls these 'meso-level' decisions. The ability to control the agenda is also the ability to affect the final policy outcome, even if the power of enactment exists elsewhere. The CAP reforms which are enacted are shaped by the reforms proposed.

These meso-level decisions are not the exclusive preserve of the Commission. There is a debate about the circumstances in which the Commission will propose a reform; that is, exercise the policy-shaping function. The decision to promulgate reform ideas, which are always in some form circulating in the Commission, must to some extent be based on a political calculation of what the CoAM will agree. Hence any explanation of the decision to propose a reform needs to specify what motivates the Commission, and what the Commission thinks motivates the Council. These will not necessarily be the same thing. Further, in the reforms of 1984 and 1988, the European Council was active in forcing a CAP reform agenda on the CoAM. The institutions framework encourages research into the perspective of different institutions on CAP reforms. The notion of process allows the conclusion that what motivated the proposal of CAP reforms by the Commission is not necessarily linked to what prompts the final enactment of reform by the CoAM or the European Council.

Bulmer and Wessels (1987) argue that the European Council has increasingly been forced into the role of 'Court of Appeal', as was the case in the reforms of 1984 and 1988, by failures in the EU decision-making system. They note the lack of a co-ordinating institution between various Councils. As public policies have become increasingly complex, they have traversed the functional divisions of the Councils; an example is the spillover of agricultural policy issues into trade, foreign affairs and budget areas. The European Council does not co-ordinate; it has no regular agenda and does not allocate responsibilities among the various Councils. Instead, the various Councils engage in a series of turf battles with each other, with the issues of greatest contention pushed up to European Council level for resolution. This is what happened in the CAP reforms of 1984 and 1988.

5.8.2 The Institutional Context of the Pressures for CAP Reform

When a factor is specified as a cause of a CAP reform it is necessary to state where and what the incidence of that factor is – in particular, how it affects the Commission, the Commission's calculations about the Council, and the Council. Further, the reform process is generally long (over a year for the reforms of 1984 and 1988) and decisions are made by different institutions at different times. Thus, even if it is claimed that the same factor affects all institutions, it affects them at different times and in different circumstances.

The institutional path starts with the Commission and ends with the CoAM. It is these two institutions which determine the characteristics of the CAP reform process and the CAP reforms. Each is considered in turn. The Commission provides part of the explanation of why the CAP is so difficult to reform: DG II (Economics) and DG XIX (Budget) have wrestled (unsuccessfully) with DG VI for control of the CAP agenda. The horizontal separation of DG VI by commodity division hampers the construction of reform proposals in that institution. Reforms are conceived in small groups and cliques away from the main policy or administrative channels of DG VI. The role of Commissioners' cabinets was a material factor in the reforms of 1984 and 1988.

The second part of the path which allows CAP reforms to be understood is an examination of the individual members of the CoAM. The third premise of the public choice paradigm stated earlier in the chapter applies here: the CoAM is pro-agriculture because its members are politicians who bring the perspective of incumbent ministers of agriculture. The influence of agricultural policy interest groups in the EU's black box is much less than the influence of this basic institutional fact. The source of explanation of CAP reform lies in why a qualified majority of members (at some time) are constrained to accept the inevitability of reform.

There are two ways to argue that the institutional context dominates interest group pressure as motivating a national government's position, with respect to a proposed CAP reform. The first emphasizes that each member of the CoAM is a member of a national government, which has to agree at an executive level a negotiating position for the minister of agriculture to adopt in the CoAM. Swinbank (1989), in a UK context, quotes Peter Walker from 1981 as saying that the Secretary of State for Agriculture never negotiates beyond his brief agreed with Cabinet (including the Chancellor of the Exchequer). Walker regarded the view of the CoAM as a group of agriculture ministers who push up farm prices, oblivious or indifferent to the effect on the EU budget, as a 'great myth'.

Walker's point is that because the position taken in the CoAM has to be agreed collectively by each national government, this will constrain the ability of the CoAM to expand the budget of the CAP. The key institutional point is that the position taken in the CoAM is that of the UK government as a whole, not MAFF alone. Farm interest groups have to compete with other ministries and their client interest groups for control of the UK government's position on the CAP.

The second argument focuses on the need for each member state to allow their representative to 'play the CoAM game'. A tough stance on CAP expenditure will tend to get out-voted. The game of the CoAM is a competition among member state agricultural ministries to maximize their view of the national agricultural interest. At the member state level of aggregation the policy is unequal; Chapter 7 examines the literature which measures each member state's net pay-off from the CAP. The result is that most member states are net beneficiaries from the CAP; only Germany and the UK have always been net losers, although they have been joined by other countries in certain years. This means that without the national interest veto being employed (by convention requiring unanimity in the CoAM on that issue), then the competition of national agricultural interests will continue unchecked. In the absence of the will to use a veto, nothing is gained by not arguing strongly for your country's agriculture. Hence, the CAP inexorably grows in terms of total budget cost (even though agriculture is shrinking as an economic sector), the number of products covered and the complexity of the rules required in its administration.

To understand why the issue of the CAP budget is not raised as being of 'vital national interest' by any member it is necessary to see the CAP as one element in a series of inter-state bargains. These bargains exist across a series of policy areas. Overall the balance of these bargains must be perceived by each member state as positive for that state to remain in the Union or at least accept the status quo. If positive, that member state will not upset that pay-off by using the national interest veto on the CAP budget, even if the overall CAP pay-off for that member state is negative. A veto used in a vital area of EU activity would quickly provoke retaliatory action in other areas. This would effectively bring the EU to a halt (more crudely, a pay-off of zero).

In this framework, for members of the CoAM to be 'pro-agriculture' can be interpreted as being willing to defend the existing CAP system. This is the baseline against which any incumbent measures success at the job. To allow reform would be a surrender of what some earlier minister of agriculture (or themselves) had negotiated for their nation's farm interest. The surrender of the status quo has a political cost. This defensive posture means that the CoAM has been characterized as 'myopic'. Preserving the CAP and their nation's pay-off from it in the short term is the negotiating stance of each member of the CoAM. This argument provides a way of answering the question of why the CAP is so difficult to reform.

5.8.3 Potential Weaknesses of the Institutions Framework

Three main weaknesses in the institutions framework can be identified. The first relates to the institutions framework's downplaying of the role of both COPA and national farm unions in the CAP reform process. It is noted that COPA has a relationship with the Commission, and the national farm unions have a relationship with their respective national agriculture ministries in the administration of the existing agricultural policy. There is an argument that this shared administrative agenda feeds into the CAP reform process. It is in these consultations that the problems of the existing CAP are identified and passed into the CAP reform agenda.

The second potential weakness is the underestimation of the influence of agribusiness interest groups. These groups will always tend to be less obvious or public in their political pressure than farmers. This is because they are not organizations with a large and often noisy membership like farm unions. However, the economic benefits that, for example, food processors and fertilizer manufacturers gain from the operation of agricultural support suggest that agribusiness ought to have an interest in influencing the direction of CAP reform. The institutions framework does not explicitly account for this, but admits that the agribusiness interest is another factor that could influence the positions of national agriculture ministers in the CoAM. This influence would be independent of any explicit lobbying by agribusiness interest groups.

The third potential weakness of the institutions framework concerns the role of individuals within institutions. The framework holds that it is the institutional context which affects an individual's perspective, their objectives and their actions in the CAP reform process. The framework does not provide any guidance on the effectiveness of individuals' actions. Even if the overall strategy and actions of an individual are determined by their institutional context, it may be that some individuals' actions will be more effective than others'. The effectiveness of an individual's action is a function of a wide range of factors which may be grouped under the category of political skills. In particular, political leadership is an issue in CAP reform processes. This is something revisited in Chapter 6. The institutions framework holds that individuals affect the reform process. However, that effect is determined by a range of influences, not just the individual's institutional context.

5.9 THE SELECTION OF THE INSTITUTIONS FRAMEWORK IN THIS BOOK

This section sets out the reasons for declaring an initial bias towards the institutions framework as the most convincing of the three rival frameworks to use in Part II of the book and confront with the evidence of the MacSharry case study. This declared bias is made on the basis that the institutions framework provides the most cogent interpretation of the histories of the reforms of 1984 and 1988. Although Petit *et al.* (1987) and Moyer and Josling (1990) respectively wrote the histories of the 1984 milk quota reforms and the 1988 stabilizer reforms within the prominent players framework, it is the contention of this book that the evidence they collected could have been better interpreted through the institutions framework. There are three parts to this claim.

The first part of the claim is in terms of the interest groups framework: the accounts of the reforms of 1984 and 1988 contain no evidence of a shift in the balance of interest group power which created a pressure in the CAP decision-making system for reform. The second part states that the histories of the reforms of the CAP in the 1980s do not give the impression of the operation of policy communities at national or supranational level. Third, the CAP reforms of 1984 and 1988 appear to be the product of a system of institutions in conflict in an atmosphere of political crisis.

The applicability of the interest groups framework depends on evidence of a changing balance of interest group power in the CAP decision-making system. An interest group becomes more powerful when either the payoffs to its lobbying activities increase or it enjoys more political resources with which to lobby. Chapter 3, setting out the histories of the CAP reforms of 1984 and 1988, provides no evidence of either of these trends in interest group power. It appears that interest groups were not a significant factor in either the types of reforms proposed or the timing of the proposals.

Policy communities, highlighted by the prominent players framework, were not a feature of either CAP reform process at an EU institutions level. There was some limited evidence of their existence in certain member states, but an insufficient amount to support a claim that they were crucial factors in driving the reform processes in 1984 or 1988.

The more cogent analytical framework for the reforms of 1984 and 1988 is the institutions framework. It describes the black box of the CAP decision-making system as operating in a way resembling that in which it appears to have operated in the reforms of 1984 and 1988 as described in Chapter 3. This can be described as follows. Each member of the CoAM has a distinct view of their nation's agricultural interest. This is generally unchanging over time. However, the view of the minister as to how this interest is best furthered can change according to circumstances in the CoAM. It is the estimation of the political costs of agreeing CAP reform by a national agricultural minister which determines each member state's position in the CoAM.

The CoAM is a series of competing national interests in the CAP. The dynamics of the CoAM affect the judgement of each member about whether to agree to reform – specifically, whether the proposal will be agreed without his or her vote, and, on that basis, whether the political costs of not agreeing reform are greater than those of agreeing to a CAP reform.

The CoAM exists in conflict with other institutions in the black box, most notably the Commission's ambitions for CAP reform. It is in the CoAM–Commission relationship that the key tension of the CAP reform process exists. Without the Commission's support, CAP reform proposals require unanimity, and agreement in the CoAM is consequently that much harder to reach. When the CoAM's incapacity to reach an outcome on CAP reform affects other aspects of the EU, notably the EU budget, the European Council may come involved. It is only here that the institutional stranglehold of agricultural ministers on the agreement of CAP reforms is broken. The outcome of the CAP reform process can be reached by heads of state or government weighing CAP reforms against their nation's wider interest in the EU.

5.10 CONCLUSIONS

This chapter has set out the public choice paradigm of decision-making and argued that it is appropriate to employ it in the construction of a positive view of the CAP decision-making system. The essential characteristics of a public choice approach were set out as follows. The understanding of any public policy decision must be grounded on an account of the individual actions involved in that decision. Individual agents are, to a large extent, rational and their preferences and view of the decision-making system are to a large extent conditioned by their position within that system. Different views of the structure of decision-making systems produce a diversity of models within the public choice paradigm. The role of individuals within the structure of the CAP decision-making system is a theme throughout this book, and is drawn together and considered in Section 9.4.1.

Three frameworks of the CAP decision-making system have been set up: the interest groups, the prominent players and the institutions. The institutions framework has been selected as the most appropriate to use in building the MacSharry case study in Part II of this work. The

CoAM–Commission relationship is highlighted by this framework as the key factor in influencing when, why and how the CAP is reformed. The interest groups and prominent players frameworks both shed light on different parts of the CAP decision-making system. These have some influence on the CAP reform process. The network of agricultural policy interest groups and farm ministries can affect national governments' position during reform negotiations. Similarly, the lobbying activities of agricultural policy interest groups cannot be discounted as a factor. It is the contention of this work, however, that the germane elements of the policy processes associated with the CAP reforms of 1984 and 1988 are more obviously captured by the institutions framework.

NOTE

1. Political scientists tend to use the term 'state' in such discussions to preserve a distinction between the state and the government. The formal models in the first strand of the public choice literature commonly use the term 'government'. The two terms are used interchangeably in this chapter.

Chapter 6

The Policy Process of the MacSharry Reforms

Chapter 4 described certain postulated causes of the MacSharry reforms. It examined how the CAP was operating at the time, noting that problems with the status quo have been central in previous reforms. Further, the Uruguay Round had been part of the agenda of EU agricultural policymakers since 1986. The organization of the world trading regime for agriculture and the CAP's part in it was an outstanding issue at the collapse of the Uruguay Round at Brussels in December 1990. The oilseeds dispute was a separate, but related, international trading issue.

This chapter details the links between these circumstances, nominated as causes, and the enactment of the MacSharry reforms in May 1992. These links form a chain – the policy process. As described in the introductory chapter, any decision to reform the CAP is the outcome of a policy process. The policy process exists within the black box of the CAP decision-making system. This chapter gives a perspective on the operation of the black box in an actual observed policy outcome.

The reform process ended with the enactment of Council Regulation (EEC) no. 1765/92 30/6/92. It began with the decision by the MacSharry reform team to bring forward CAP reform proposals to the Commission. This first event cannot be dated accurately. December 1990 is taken as a proxy, because after the collapse of the Uruguay Round at Brussels on the 6th, the information that MacSharry intended to bring forward CAP reform proposals gradually came into the public domain during the following week (*Agence Europe*, 11 and 12 December 1990 and *AgraEurope*, 14 December 1990). However, as we shall see, Ray MacSharry's intention of bringing forward CAP reform proposals at some stage in his tenure as Agriculture Commissioner had existed well before then.

The rest of the chapter is arranged around a further ten events. These define stages in the MacSharry reform process and form Sections 6.2–6.11.

Section 6.12 concludes with a discussion of the main causal links in the policy process of the MacSharry reforms.

6.1 EVENT 1: DECISION TO BRING FORWARD REFORM PROPOSALS WITHIN THE COMMISSION

This section addresses the question of why there is a reform process at all. To this end, it is structured around three questions: which individual agents were involved in the decision?; what motivated their general ambition for CAP reform?; and why was the formal process triggered in December 1990? These questions will help to identify the links operating at different times and at different levels in the reform process.

6.1.1 The Individuals Involved in the Decision

A team within the Commission had been considering plans for a reform of the CAP since the beginning of the second Delors presidency and the management of the agriculture portfolio by MacSharry. The official starting date of this presidency was 6 January 1989. The team was composed of MacSharry, Patrick Hennessy (Deputy Chef de Cabinet of the MacSharry cabinet), Guy Legras (Director-General of DG VI) and Demarty (the agricultural policy specialist in the Delors cabinet). The MacSharry cabinet met separately. In addition to Hennessy, it comprised Verstaylen and MacDonagh, who worked on the construction of the reform proposals, and Minch, who specialized in Uruguay Round issues. Larkin, the Chef de Cabinet of MacSharry, was a specialist in structural policy and not directly involved in the construction of the MacSharry reforms.

The first element involved in the existence of an ambition for CAP reform within part of the Commission in 1990 was the selection of MacSharry for the agriculture portfolio in 1989. Delors had decided even by early 1989 that the CAP would require reform beyond the stabilizer regime. Delors's view was that the stabilizer regime would be only a medium-term measure to control the rate of increase of the CAP's budget costs. If his long-term ambitions for the EU (e.g. EMU, and large social and regional policies) were to be realized, he believed that further reforms of the CAP would be necessary. Delors considered MacSharry a suitable proponent of such reforms for the following reasons. MacSharry had an agricultural background and knew how national farm ministries and farm lobbies viewed the CAP. In addition, he had a definite personal commitment to the CAP. Further, he was the Commissioner from a country which historically has been a large beneficiary of the policy. It is possible to trace the MacSharry reform process back to this choice by Delors. MacSharry's political leadership will be a theme throughout this chapter. At various times it is asserted as causal factor in the reform process. According to John Gummer, 'Without MacSharry's drive and determination it must be doubtful whether the reforms would have succeeded, at least in the form in which they eventually emerged.'

The subsequent disagreements between MacSharry and Delors over the substance of the reform proposals and tactics of leading the process are detailed in this chapter. MacSharry's tenure as Agriculture Commissioner does not conform to the Delors–commissioner relationships characterized for the first Delors presidency (1985–1989) in recent studies of the Delors presidencies (Grant, 1994; Ross, 1994, 1995). Ross (1995) admits that a number of commissioners of the second Delors presidency existed as counter-examples to his model of presidential control of the agenda and the implementation of EU public policies. MacSharry was a high-profile counter-example.

6.1.2 The Delors–MacSharry Relationship

It is important to establish some points about the relationship between Delors and MacSharry early in this chapter. The argument will be made that Delors was an important factor in the enactment of CAP reform, in the background during the reform process. My research material and secondary sources focus on the MacSharry factor; MacSharry and his team conducted the CAP reform campaign. Delors was involved in the general ambition of a team within the Commission to reform the CAP.

Ross (1995) sets up a standard model of Delors–commissioner relationships. The horizontal relationship between a commissioner and the services of the areas of his or her portfolio can be variable. The variation depends on the commissioner's ability to give a political and administrative lead. Delors made this relationship even more potentially volatile; he had a *strategic* view across a whole range of policy areas, and along with Pascal Lamy (Chef de Cabinet) built an inner circle of trusted Director-Generals. This would allow the Delors cabinet to 'reach around' a Commissioner to his or her policy area. Delors would set out the main 'policy lines' (Ross, 1994) in any area and it would be the responsibility of each member of his cabinet to work on the implementation of these policy lines. Lamy in Ross (1995) called it *autogestion militaire*.

Ross (1995) notes a difference between the first and the second Delors' presidencies. After 1989, the number of commissioners who 'owed' Delors something on a personal scale dropped. The period 1985–1989 can be characterized as successful Delors initiatives creating pay-offs for certain Commission services and the associated commissioner. Delors would allow the credit to remain there in return for support for other presidential plans.

MacSharry inherited a high-profile portfolio and had no direct personal allegiance to Delors. He had an effective cabinet with ideas and was himself a shrewd political operator. The reach-around tactic, of going to the services of a commissioner behind the commissioner's back, was not possible with MacSharry, because he brought with him, and developed, his own agenda.

After the collapse of the Uruguay Round in December 1990, Delors and MacSharry agreed on the need for CAP reform, agreed that this reform should come before the Uruguay Round, and agreed on the main substance of the proposed reforms. They shared a sufficiently close vision of the CAP that there was no battle for control of agricultural policy. Delors's enthusiasm for reform of the CAP was one factor in the College's adopting proposals for its reform and therefore in the initiation of a reform process. The factor of Delors's and his cabinet's skills at Commission politics is described under the 'Event 2' heading (Section 6.2).

During 1991–1993, there were a number of instances when the relationship between MacSharry and Delors soured and there was disagreement over agricultural policy. The most serious disagreement concerned the conduct of the GATT negotiations (the agriculture section of which was MacSharry's responsibility). The Uruguay Round, the oilseeds dispute and the circumstances which precipitated MacSharry's resignation in November 1992 are covered in Chapter 7. There was also the dispute over the guideline of the 1991 farm budget noted in event 4 (Section 6.4) and in Chapter 4.

The reform team reflected MacSharry's desire for small, informal working groups. This helped to maintain a degree of secrecy throughout the development of reform ideas. It is clear that few of the commodity division heads of DG VI knew that reform proposals were being prepared in 1989/90. Details emerged into the public domain only after MacSharry announced that reform proposals would be brought before the College. Legras also favoured this working style.

The preamble to this chapter explained how the institutions framework established in Chapter 5 gives a role to political leadership in the explanation of the output of political systems. The question of the political leadership of Commissioner MacSharry being a causal factor in the CAP reform process is considered at different points in this chapter. The term 'political leadership' is sufficiently broad to cover a range of the beliefs and actions of MacSharry and his team at various points in the reform process. As such, there is no single political leadership link in the chain of the reform process. Factors covered by the term are different and operated at different times.

MacSharry brought personal views and convictions about agricultural policy to the position of Agriculture Commissioner. These affected his, and the reform team's, reading of the recent history of the CAP before 1990. MacSharry's view of the CAP in 1990 rested on two instincts. The first was the conviction that the EU cereals regime was more vulnerable in international trade negotiations than the USA's regime. The EU would face strong demands for large reductions in the internal prices of its CAP regimes. The point was that the CAP involved explicit import barriers and high consumer as well as producer prices, whereas the USA deficiency payments system avoided import controls and kept prices to consumers low. Also, the rapid import growth of cereal substitutes coming in under zero tariffs for oilseeds would be difficult to stem by increased protection in an atmosphere of agreeing reductions in total agricultural support. The rebalancing elements of the EU's Uruguay Round submissions were controversial and occupied much negotiating time and capital with the USA delegation.

The second instinct was that the central problem of the CAP was described by the statistic that 80% of EU farm spending went to only 20% of farmers. MacSharry used to spend time studying tables showing the distribution of the benefits of the CAP at an EU and member state level. The MacSharry interview provided a number of anecdotes about ships sailing around the North Sea because there was nowhere else to store the grain and about the profit levels of storage companies.

MacSharry's instinct was that the 80/20 statistic illustrated how the CAP, as it had operated, rewarded production. The link of support spending to production had two effects. First, it was unfair because the most support was received by the most efficient farms, those that needed it least. Second, it had resulted in expensive surpluses. MacSharry had been aware of proposed income support schemes for a number of years. They had existed in agricultural policy circles and academia for almost 20 years (for example Josling, 1973). It was an act of political leadership to set a reform team in DG VI to work on CAP reform proposals which moved away from high support prices as the main channel of support for farmers.

The modulation of the proposed DIPs was an element which was personal to MacSharry. He viewed a function of the CAP as keeping the *maximum* number of the EU's 10 million uneconomic farms in existence. Modulation was an attempt to redistribute support away from the most productive 20% of farmers, achieving the twin objectives of redressing the imbalance in the allocation of support and the controlling of surpluses.

His personal philosophy was described as a 'very west of Ireland view' and reflecting MacSharry's cattle-farming background. That support was just as acceptable through DIPs as through a price support system, and second, it was common sense that to control surpluses required a control of the most productive farms.

6.1.3 The MacSharry Reform Team's Reading of the History of the CAP

This chapter starts the history of the MacSharry reforms from the introduction of the stabilizer regime in 1988. The discussions of the reform team in 1989 and 1990 of the general requirements for a CAP reform started their history from this point. It is important to note how the reform team understood the stabilizer regime; it is part of the explanation of why the status quo was rejected and a general ambition to reform the CAP was converted into concrete proposals to bring to the Commission.

The stabilizer regime provided for price cuts in the case of cereals production exceeding the MGQ for that year's cereal production and introduced a voluntary set-aside scheme. The senior officials of DG VI discerned divergent views among the member states as to the role of voluntary set-aside. It was possible to characterize two extremes, specifically the UK view and the German view. The UK considered that the voluntary set-aside scheme was a safety valve to mitigate the effects on farmers' incomes of the inevitable price cuts that would happen under the stabilizer regime. The German view was that the voluntary set-aside scheme represented a structural measure to avoid price cuts.

Legras commissioned an internal DG VI review of the stabilizer regime. The report was carried out by the Special Studies unit of DG VI, headed at the time by Dirk Ahner. It was completed and reported to the reform team in July 1990. The report was constructed in secret, like the deliberations of the reform team. Few of the commodity division heads of DG VI knew about either. However, its contents drew extensively on data and reports from the commodity divisions. The results of the report are hinted at in *AgraEurope* of 14 July 1990.

The report's conclusion was that the stabilizer regime was not having the intended effects. Specifically, it had not arrested the growth in surpluses and budget costs which had promoted its introduction (see Chapter 3). The background to the report was mounting problems in the cereals and livestock sector. There had been heavy and expensive intervention in the sheepmeat regime. Beef surpluses had grown because of the BSE scare in the UK. This problem was compounded later in 1990 as the Gulf War cut exports.

Chapter 4 details the problems agreeing the 1990/91 price package. Although automatic price cuts were triggered, there was considerable debate over the whole operation of the stabilizer regime. The reform team considered there to be a shift in the attitude of the CoAM towards the stabilizer regime. Importantly, they detected a movement in the Dutch position from the UK to the German view of set-asides and price cuts: that the set-aside part of the 1988 reforms was a supply control measure designed to avoid the automatic, uncompensated price cuts then being imposed by the stabilizer. Although an importer of cereals, the Dutch government reacted against the reduction in the ability of the CoAM to adjust CAP support prices annually and a perceived agglomeration of power in the Commission. The shift meant that the stabilizer regime was open to question and reform in the CoAM. The Dutch government had not been a traditional supporter of high prices in the CAP as Dutch agriculture has been among the most efficient in the EU. This change in the beliefs of the reform team about the CoAM's attitude to proposals of CAP reform was a factor in the decision actually to bring forward concrete plans.

The scheduled conclusion of the Uruguay Round for December 1990 meant that the Commission required a negotiating stance. De Zeeuw had set 1 October as the deadline for the submission of final offers. As discussed earlier, this need for the CoAM to agree some mandate dominated its agenda in the autumn of 1990. This requirement was a causal factor in changing the nature of the debate about and the agenda for CAP reform.

After seven meetings and the political turmoil, the CoAM agreed a negotiating mandate which included an offer of a 30% reduction in some aggregate measure of domestic agricultural support over the period 1986–1995. This offer was tabled by the EU on 7 November, over one month late. The Commission's proposals to the Council reflected a desire to counter the zero option tabled by the USA. The aim of the final offer in the Uruguay Round was 'credibility': making it clear that the EU was serious about agriculture being part of the Uruguay Round negotiations.

The eventual agreement of this offer, from the perspective of the MacSharry reform team, meant that arguments over the automatic 3% cuts of the stabilizer regime had become irrelevant. The view of MacSharry and his advisers was that these automatic price cuts would not have delivered the 30% reduction in domestic support offered in the agriculture part of the Uruguay Round. The international arena had imposed a CAP reform agenda of substantial price cuts and the debate had shifted to compensation for those cuts.

The Uruguay Round collapsed in Brussels on the night of 6 December 1990. The apparent chance of progress on the basis of a paper by Hellström, chairman of the Agriculture Negotiating Group at ministerial level, was based on the perception that the EU Commission was willing to use its 7 November offer as a starting-point for negotiations (i.e. go beyond it for final agreement). This perception had faded by the evening of 6 December, when MacSharry stated that he could negotiate only to the limit of the 7 November offer and no further. The Brussels meeting broke up and the Uruguay Round was suspended. The USA and the Cairns Group accused the EU of intransigence in admitting GATT constraints on the operation of the CAP, and blamed the breakdown of the round on this intransigence in agriculture. During the following week, Friday 7 to Friday 14 December 1990, the MacSharry reform team began to prepare to bring forward its reform proposals, and the press carried reports that the Commission was studying plans for a radical reform of the CAP. It is the link between these two events which needs to be elaborated to establish the factors involved in the trigger of the MacSharry reform process.

6.1.4 The Link between the Collapse of the Uruguay Round and the Promulgation of the MacSharry Reform Plans

The official Commission line was that there was no link between the collapse of the Uruguay Round and the start of the CAP reform process. It was admitted that the agreement of CAP reform would have effects on international negotiations, but the claim that this meant that the collapse of the round had triggered the introduction of CAP reform proposals was vehemently denied. This official interpretation of the link between the two events does not stand up to scrutiny, and was used by MacSharry and his reform team as part of their publicity campaign to the effect that CAP reform was being proposed for entirely domestic reasons.

The reason why the official interpretation of the link does not stand up to scrutiny is that it ignores the institutional rule involved; MacSharry was responsible both for any proposals for CAP reform and for the agriculture part of the Uruguay Round negotiations. In trying to achieve these two objectives, MacSharry had two constituencies. He could not ignore the fact that the issue of CAP reform existed in the Uruguay Round. Even if CAP reform and the round were not linked for the EU, they were for the EU's negotiating partners. It is for this reason that the official interpretation is inadequate. MacSharry's ability to make progress in the Uruguay Round depended on his ability to make progress on CAP reform.

MacSharry links the two events, which means that understanding his political leadership is central to understanding the link between the collapse of the round and the start of the CAP reform process which led to the MacSharry reforms. There are two interpretations of the political leadership of MacSharry and his reform team over this period.

The first interpretation has been suggested by Tangermann (1996), who takes the position that MacSharry, in eventually achieving CAP reform and an agreement in the Uruguay Round, pulled off a 'political masterstroke'. Specifically, MacSharry deliberately avoided agreement at Brussels in December 1990 in order to create the circumstances for a reform of the CAP and an eventual agreement in the Uruguay Round. The second interpretation (and the one advocated in this work) of the link between the temporary collapse of the round and the start of CAP reform proceedings suggests that MacSharry's leadership strategy was very much limited by the circumstances of the middle of 1990. These circumstances forced him and his team into rescuing the EU's credibility in the Uruguay Round, and ensuring that the negotiations were able to be completed (albeit some time after the original deadline). The circumstances of mid-1990 were, to an extent, the result of a lack of political leadership by MacSharry.

The Tangermann (1996) interpretation makes the following steps:

1. The 7 November offer of the EU, if it had been accepted at Brussels by the USA and the Cairns Group, implied CAP reform.

Details of the 7 November offer are provided in Chapter 4. The argument for this step is that the stabilizer regime as it was then operating would not have produced sufficient price cuts or limited production enough to meet the commitments of this 7 November offer. The argument is speculative; as discussed earlier, there were no concrete export subsidy commitments in this offer.

2. The sequence of Uruguay Round agreement then CAP reform was not politically feasible.

To reform the CAP under the pressure of a legally binding international agreement was simply politically impossible. Tangermann (1996) invites the reader to 'imagine the political turmoil this would have caused in the EU'.

3. Hence, the MacSharry reform team knew that CAP reform had to come before the Uruguay Round agreement. Two things, (4) and (5), follow from this.
4. MacSharry could not have accepted agreement in the Uruguay Round at Brussels in December 1990. He and his team knew that the round, or at least the agriculture part, would have to be delayed.
5. Yet, given (4), he had to persuade the negotiating factions of the Uruguay Round, principally the USA and the Cairns Group, to delay the round rather than give up on it entirely. For this it was necessary to persuade the USA and the Cairns Group of the credibility of the EU's desire to reduce domestic agricultural support levels.
6. The manner of the collapse of the Uruguay Round at Brussels in December 1990 was 'politically masterminded' by MacSharry.

The manner of the collapse was crucial to MacSharry's political leadership strategy. He managed to convince the EU's negotiating partners that the EU was serious about reductions in domestic agricultural support (and hence CAP reform), and that agreement could be reached at some point in the future, while at the same time appearing to the domestic constituency to be defending the CAP against international pressures by pulling the Uruguay Round down. MacSharry maintained credibility in the international arena, while building credibility in the domestic arena, to the extent where he was able to claim that CAP reform proposals were being brought forward for purely domestic reasons.
7. The fact that CAP reform was enacted in May 1992 and the agriculture part of the Uruguay Round was concluded at Blair House, near Washington, DC, in November 1992 (see Chapter 7) shows that MacSharry's actions at Brussels constituted strong political leadership. He deliberately avoided agreement at Brussels in such a way as to allow the Uruguay Round to set the circumstances for domestic CAP reform, but not obviously cause those reforms. Simultaneously, the round was kept alive, and MacSharry was able to show that the EU was serious about an internationally acceptable CAP.

The results of my fieldwork for this book suggest a second interpretation of the chronology of events from the middle of 1990 to the initiation of CAP reform proposals in the Commission in December 1990. This interpretation can be set out in the following steps:
1. There had been a team planning CAP reform in the Commission since MacSharry began in the Commission in January 1989.

Fieldwork interviews provided this information and also evidence for the following step (2) in this second interpretation of the link between the collapse of the round at Brussels in December 1990 and the initiation of CAP reform proceedings in the Commission in that same month.
2. The MacSharry reform team believed that the order of the two events had to be CAP reform first, then Uruguay Round agreement. CAP reform

was necessary for the completion of the Uruguay Round. This was because (i) there was limited chance of agreement with the USA and the Cairns Group because the EU could not move far enough, given the present CAP system, and there was not enough potential area for compromise, and (ii) it was a politically difficult sequence from Uruguay Round agreement to CAP reform.

3. The MacSharry reform team failed to get a CAP reform agreed before the middle of 1990.

This may be attributed to poor political leadership by MacSharry, or alternatively to a lack of time between MacSharry's beginning in the Commission (January 1989) and the middle of 1990 in which to achieve CAP reform.

The planning of reform proposals was in progress during 1990 and it was a deliberate decision by MacSharry not to make the development of Commission thinking public during the first eleven months of 1990. *Agence Europe* (11/12 December 1990) reported that the promulgation of developmental reform proposals was delayed in 1990 in order to avoid affecting MacSharry's negotiating position in the Uruguay Round. MacSharry confirmed this, stating that his position in the round would have would have been weakened by 'ill-timed announcements'. His team confirm that the proposals were inchoate at the time the EU required a negotiating stance in what was supposed to be the conclusion of the Uruguay Round (mid- to late 1990). Ross (1995) quotes an interview with Demarty in which Delors saw the collapse of the round as a blessing in disguise; it allowed time for the reform proposals to be more fully completed.

The judgement of MacSharry's political leadership rests on the extent to which he was able to influence the circumstances of the middle of 1990; that is, the EU having no CAP reform agreed and seemingly having no prospect of CAP reform being agreed.

Given the enactment of the stabilizer regime in 1988, and the usual cycle between CAP reforms, it is a fair point to argue that MacSharry could not have developed CAP reform plans any more quickly than he did. However, the question remains whether he should have made public the kind of ideas that were being considered by the Commission during 1990 to try to help the EU's negotiating position in the Uruguay Round. Instead, there was step (4) of this alternative interpretation of the link between the collapse of the round and start of the MacSharry reform proceedings.

4. Step (3) meant that the EU was forced into a rescue operation to save the Uruguay Round. The MacSharry reform team had to spend all their time battling to achieve the 7 November offer.

5. The MacSharry reform team believed that the 7 November offer was incompatible with the CAP as it existed. This analysis rested on two beliefs. The first was that the stabilizer regime would not have produced sufficient reduction in domestic support in order to meet the targets of that offer. Second, the uncompensated automatic price cuts of the stabilizer regime were increasingly politically difficult. The 1990/91 price package had been problematic. There was a key movement in the CoAM in the position of the Dutch government. The Dutch position was not usually associated with a commitment to high support prices; in the stabilizer reform debate the Dutch had broadly taken the UK view of set-aside (see above). However, the second year of uncompensated price cuts had been politically difficult enough for the Dutch to move their position towards that of the German government, namely, that the CAP needed to be changed in a way which would avoid annual uncompensated price cuts. Thus, the analysis of the MacSharry reform team was that the stabilizer regime as it was then operating was too tough for the CoAM; there was no prospect of tightening the system in order to meet the requirements of the 7 November offer. The operation of the stabilizer regime and the automatic uncompensated price cuts that it imposed are a central factor in why the CoAM eventually agreed to the MacSharry reforms in May 1992.

6. The MacSharry reform team thought that the 7 November offer would not be enough to get agreement, but hoped it would be enough to keep the Uruguay Round alive.

MacSharry admitted that the offer made by the EU on 7 November 1990 acted as a constraint on the rest of the Uruguay Round negotiations. His judgement was that an eventual agreement in the negotiations would depend on the EU validating that offer. After Brussels in December 1990, MacSharry elaborated his attitude to the connection between CAP reform and the Uruguay Round in *Agence Europe* (9 January 1991): CAP reform would help to 'strengthen the Community's credibility on the international scene by affirming our will to strengthen our farm policy' in the GATT negotiations. This leads to step (7) in the alternative interpretation.

7. The promulgation of reform proposals so soon after the collapse of the Uruguay Round at Brussels in December 1990 was part of an attempt to build EU credibility on agricultural support reduction. The aim was to avoid the failure to reach agreement at Brussels, and irrevocable damage to the Uruguay Round.

The DG VI's GATT team and the MacSharry reform team met jointly a number of times during the week 8–14 December 1990. MacSharry was the common element of the two teams. They had the problem of a collapsed Uruguay Round in which the EU's offers lacked credibility with the USA and the Cairns Group, and a stabilizer regime which they believed was not controlling surpluses and, in addition, was producing politically unacceptable price cuts. The view of the GATT team was that the USA wanted to end export refunds, but keep its deficiency payments system. It was MacSharry's decision to proceed with proposals for a CAP reform based on substantial price cuts and compensatory payments.

8. That decision concluded a successful holding operation by MacSharry; the Uruguay Round was kept alive, something that was in no way inevitable given the events and circumstances of the second half of 1990. However, those circumstances were in part influenced by the failure to achieve CAP reform, or at least get reform on the EU's agenda before the original deadline of the round.

MacSharry knew that the CoAM was responsible both for CAP reform and agreeing the Uruguay Round negotiating mandate. He also knew that there was a basic political constraint that the two issues could not be publicly linked; that is, be discussed as part of the same agenda. Given this, MacSharry should have got some kind of CAP reform agenda agreed (even if not formally enacted in detail) before the scheduled final stages of the Uruguay Round in the second half of 1990. There was no guarantee that the round would survive a collapse at Brussels, so a strong political leadership strategy would have been based on the view that the round was going to be completed at that time. It was weak political leadership to allow the circumstances of the middle of 1990 to develop as they did.

Under MacSharry's political leadership, the EU did agree to reform the CAP in May 1992 and did conclude the agriculture part of the Uruguay Round in November 1992. Chapter 7 compares De Zeeuw's paper of 1990 with Dunkel's Draft Final Act of late 1991. It is speculated that one factor why the latter rather than the former became the basis of the Uruguay Round agriculture agreement was that the prospect of CAP reform existed in late 1991 but not in the summer of 1990.

There are two main differences between the two interpretations of the collapse of the Uruguay Round and the beginning of CAP reform proceedings. The first is that the Tangermann thesis ignores the fact that MacSharry and Delors had a team working on CAP reform for almost 2 years before December 1990. The failure to get CAP reform on to the CoAM agenda was a failure of political leadership. The circumstances which forced MacSharry into the 'masterstroke' (according to the Tangermann thesis) were at least in part affected by MacSharry himself.

Second, the Tangermann thesis about MacSharry politically masterminding CAP reform stretches credulity. There was no guarantee that the Uruguay Round would restart after December 1990, so it would have been a reckless strategy to deliberately collapse it. The alternative interpretation suggests that the MacSharry team did their best to achieve agreement at Brussels; indeed, the premise on which their estimation that CAP reform had to come before the Uruguay Round was that what the EU could offer with the stabilizer regime would not be sufficient for agreement. In interviews, members of the MacSharry reform team said they would have been very pleased to get agreement on the basis of the 7 November offer. They would have limited the international constraints on the future development of the CAP and could have claimed that the USA had accepted their proposals for a post-Uruguay Round agricultural trading system.

After the decision to proceed with proposals for CAP reform had been taken in December 1990, the work on different CAP regimes by the reform team was being pulled together with 'hard' figures and a full quantitative assessment of the effects of such a reform of the CAP. *AgraEurope* of 18 January 1991 printed selected texts from a copy of the initial reform document which it had obtained. The proposals were incomplete in the sense that the politics had not been added to the reform proposal. It was never intended for the public domain or even the College of the Commission. The research for this book did not produce a convincing account of why it was leaked.

6.2 EVENT 2: ADOPTION OF COM (91) 100 BY THE COMMISSION AND INITIAL REACTION IN THE PUBLIC DOMAIN

Event 2 forms an important chain in the CAP reform process. As highlighted in this section, the reform proposals progressed from the initial stage of small-team discussions in DG VI (see Event 1 above and the *AgraEurope* leaked document) to agreement by the College on 31 January 1991.

The reform team used the tactic of informal and individual consultation within the Commission; individual commissioners were targeted on a bilateral basis for discussion and negotiation as opposed to the regular channel of special chefs, then chefs' meetings followed by the College of the Commission. This was an example of MacSharry and Delors working together.

The highly irregular tactic of using a special weekend seminar of the full College replaced the usual procedure. By routeing the proposals so as to avoid the cabinets, the risk of national interests (that is, member states' governments) becoming involved and using amendments and other tactics to sabotage the project was minimized.

Members of the Brittan and MacSharry cabinets were willing to admit how explicit such lobbying was (despite the Treaty of Rome's view of commissioners). The seminar on 20 January was important in getting the main parts of COM (91) 100 (Commission of the European Communities, 1991a) agreed so soon afterwards (31 January). Only two College meetings were held to discuss CAP reform before the special weekend seminar (on 4 and 9 January). At neither of these was an official document presented.

Another effect of the weekend seminar tactic was to insulate the debate from the farm lobbies. The COPA presidium expressed 'concern and reluctance' concerning possible Commission reform proposals to change the CAP from a system of price support towards DIPs. This was in reaction to press reports; COPA did not have access to any of the detailed proposals under discussion in the Commission. Aides at COPA confirm that they had been unaware that a team in the Commission had been constructing reform plans for the CAP, including DIPs, through 1990. Yverneau, COPA president, publicly expressed concern over the direction of the reform plans after an audience with MacSharry (*Agence Europe*, 11 January 1991). COPA effectively was in the public domain in terms of understanding the progress of Commission and Council thinking during the MacSharry reform process; that is, they did not have 'insider status' (Grant, 1989). The MacSharry cabinet thought the views expressed by the COPA presidium were reactionary and conservative; the reform proposals shattered the hard-won consensus that COPA had reached on the previous reform of the CAP in 1988.

The weekend seminar tactic was decided by the reform team. For the institutional procedure of the College of the Commission to be circumvented required the input of Delors. Delors was personally involved in this part of the reform process. The tactic is an example of the ability of the Delors cabinet to impose a view on the Commission. The weekend seminar meant that no other commissioner could provide an alternative vision because their cabinet did not have time to agree a response. This tactic was typical of those employed by the Delors cabinet in its irredentist activities in other policy areas, as described by Ross (1995).

The weekend seminar tactic changed the institutional path of the MacSharry reforms. It is necessary to judge whether this was a factor in the reform process. The leaked *AgraEurope* document provides a starting-point to the assessment of the effect of the weekend seminar tactic on the MacSharry reform process. No definite evidence was found that this was a deliberate leak by the MacSharry reform team. The proposals were where the reform team's thinking had reached as the Uruguay Round collapsed; that is, before they had been presented to the College or individual commissioners. Only members of the reform team knew about the details at this stage. Hence these were the initial MacSharry reform proposals before the politics within the Commission had been added. A comparison of this document and COM (91) 100 (what the College eventually agreed) is provided at the end of this section.

The main points of this initial reform document are as follows: in the cereals sector the target and intervention prices were to be cut by the end of the transition period to ECU 100 and ECU 90 per tonne compared with the then current levels of ECU 220 and ECU 165 per tonne. Direct income compensatory payments were to be introduced. The levels of compensation would be modulated on holdings greater than 30 ha; that is payments would be reduced by 0% for holdings up to 30 ha, 25% on the next 50 ha and by 35% on anything above this. Compulsory set-aside requirements were to be introduced; again these were to be modulated, the set-aside requirement being 0% of the first 30 ha, 25% of the next 50 ha and 35% for areas in production greater than 80 ha. There was no mention of compensation for set-aside.

In the dairy regime the main point was a 5% cut in the global quota. This was also to be modulated. Farms of less than 200,000 kg annual production faced no quota cut, but those with greater annual production were subject to a quota reduction of 10%. In the livestock sector, beef intervention prices were to be cut by 15% and for sheepmeat there was a limit on the reference flock for which premiums would be paid.

The informal and individual consultation between the reform team and commissioner produced alterations in the initial reform document. This document had not been presented to the College on 4 or 9 January. However, the proposals had been the subject of some lobbying by member states before the weekend seminar of 20 January. The UK's Permanent Representation to the European Union confirm that the UK government had a copy of the initial reform document in mid-December, obtained through a leak. Members of the MacSharry cabinet 'understood' that the French government also had a copy.

An article 'Brussels examines deeper cuts in EC farm support' (FT, 17 January 1991) described the preparation of a 16-page document, 'Development and future of the CAP', for the special weekend seminar on 19 and 20 January 1991. About 80% of the initial reform document survived.

Agence Europe (21/22 January 1991) and the members of the MacSharry cabinet confirm that there were two distinct sides in the College in response to the proposals presented by MacSharry on Sunday 20 January. The was the first official document presented to the College on reform of the CAP.

The commissioners of the UK, Denmark, the Netherlands and France all expressed reservations about the reform document presented. They wished to protect large farms and argued against the modulation of the compensatory DIPs. Frans Andriessen (Netherlands) and Sir Leon Brittan (UK) were considered effective, senior Commissioners by the reform team and their arguments had prominence within the College. Both supported a quick reform to facilitate a conclusion to the Uruguay Round.

The Brittan cabinet felt that the reforms would add to the budget in the short run, even if the MacSharry team could devise scenarios in which the budget cost in the medium term was reduced compared with persisting with the status quo, stabilizer regime. Andriessen emphasized the intuition that if budget problems were motivating CAP reform then the wisdom of introducing an element of support which was an addition to the budget, i.e. DIPs, must be questioned.

The opposing group of commissioners included Filippo Pandolfi (Italy), Karel Van Miert (Belgium) and Jean Dondelinger (Luxembourg). These shared the personal philosophy of MacSharry. They supported the modulation of DIPs and set-aside requirements for larger farms. The aim of the CAP, they felt, should be to keep the maximum number of uneconomic farmers on the land.

The weekend seminar finished with agreement to the principle of CAP reform but not on the concrete terms that MacSharry had submitted. The decision-making rule of the College is a simple majority vote; however, there is a strong disinclination to take a vote unless there is the prospect of a unanimous outcome. MacSharry was given a mandate to announce reform ideas to the CoAM meeting on 21 and 22 January.

AgraEurope (25 January 1991) describes the presentation of a 13-page version of the Commission weekend document (the vague and non-controversial elements) by MacSharry to the Council of Agriculture Ministers on Monday 21 January. All the figures of the original document had been removed. This presentation argued that measures to cut support should be aimed at the top 10% of farmers, who were responsible for rising surpluses and hence expenditure. The stabilizer mechanism had failed because it did not tackle what this document called the underlying problem: that CAP support is linked to production levels. The objectives of reform must be to control expenditure, increase EU competitiveness (for the inevitable increase in competition arising from GATT obligations) and to maintain rural population. This last objective is the only sensible thing for a developed rural policy, argued the proposal.

The debate in the Commission and Council over CAP reform ended its first stage when the Commission formally adopted COM (91) 100 at a full College meeting on 31 January 1991. This orientation paper encompassed CoAM and Commission views on the original and subsequent drafts of the MacSharry plan for CAP reform. This proposal was then presented to the Agriculture Council as 'Communication of the Commission to the Council: the development and future of the CAP' on 4 February 1991. It did not contain formal legislative proposals for CAP reform, but was intended as a reflections paper to stimulate an EU-wide debate.

COM (91) 100 represented a substantial watering down of the initial reform document. The division in the Commission noted after the 20 January seminar over the distinction in the levels of compensation available to large and small farmers was healed by the compromise described below. Carlo Ripa di Meana (Italy) voted against and Andriessen abstained because it did not allow enough flexibility in the Uruguay Round.

There are five significant differences between the initial reform document as revealed in *AgraEurope* (18 January 1991) and the final version of the plan adopted by the Commission, *The Development and Future of the Common Agricultural Policy*, COM (91) 100. The first is that the later document contains no mention of modulating compensatory payments (i.e. those for the cut in support prices) in the cereals and oilseeds sectors. The original contained plans for reducing the amount of compensation for the area greater than 30 ha of any holding.

Second, the two documents differ on set-aside requirements. COM (91) 100 has a flat 15% set-aside requirement with compensation paid only on the first 7.5 ha of set-aside for farms larger than 50 ha. Some modulation in the cereals sector survived the demands of Brittan, Andriessen and Christopherson. It is not clear, but there does not seem to have been any specific and separate compensation for setting aside land in the initial reform document.

The third difference is in the milk sector, where the global quota cut was settled at 3% compared with the originally proposed 5%. The distinction in the incidence of the quota cut became less severe in the later document. The specific 10% cut in quota for producers greater than 200,000 kg per annum, with no cut for those below that limit, contained in the original document did not survive. Yet some form of modulation did survive in the dairy sector and provided a target for the UK government's demands through the rest of the negotiations.

Fourth, in the beef sector the premium for male bovines was increased from ECU 120 per animal to ECU 180 per animal, paid over 3 years. The payment was still made only on the first 90 animals, thus discriminating against large producers. Further, the annual suckler cow premium was increased from ECU 40 to ECU 75 per cow with the modulation rule changed from payment on one cow per hectare of forage area to payment on the first 90 cows as above. The removal of modulation in this sector was one of the targets of the UK government's negotiating stance in the reform process. The proposals of the initial reform document for the ovine sector, limiting the premium for each producer to the first 750 animals of a flock in less favoured areas and 350 animals elsewhere, remained in place. This opposition to modulation plus the desire to reach a GATT conclusion meant that the usual UK insistence on budget stringency was not a feature of the CoAM debate on the MacSharry reform proposals.

The fifth significant difference was in the ambition. *The Bulletin of the European Communities* (no. 1/2, 1991) reports that the basic objective of COM (91) 100 was to 'enable a sufficient number of family farms to survive, thereby preserving the natural environment and contributing to rural development'. Sufficiency in this case had been substituted for 'maximum number' or 'large number' in previous drafts.

This work claims that the decision to pursue an informal consultation route to the College and eschew the normal procedures was an intentional act by MacSharry and Delors. Further, this informal consultation route was a causal factor in the MacSharry reform process; compared with the usual special chefs and chefs path, this leadership meant that COM (91) 100 was adopted more quickly and in a different form than would otherwise have been the case. These are separate claims, however the same element operates in both. The informal consultation path through the Commission meant that the intervention of national governments was minimized. National governments are the only other institutions in the CAP decision-making system which could have produced alternative overall views of CAP reform. Hence, it is their lack of input at the Commission stage of the reform process that enabled COM (91) 100 to be adopted more quickly and in the form that it actually took.

Only 2 months elapsed between the trigger of the reform process after the collapse of the Uruguay Round at Brussels on 6 and 7 December and the adoption of COM (91) 100 by the College on 31 January. The special-chefs level would have meant direct input from member state governments. Cabinet members rely heavily on their national governments for technical advice on the CAP. These inputs would have crowded the agenda and stymied the progress of the reform team's proposals at a sub-College level.

The reform proposals reached the College quickly and unscathed from competition with rival visions of CAP reform. The tactics of the reform team allowed the agreement of the principle of a CAP reform of the type envisaged by the reform team before any national government could suggest a different direction. Once the principle of a MacSharry-type reform had been agreed, the details could be agreed on a personal basis with individual commissioners. The reform team was very flexible in these negotiations; note the difference between the proposals as they started and finished in the Commission. Between 20 January and 31 January, modulation in the cereals sector was dropped (though it remained in parts of the livestock sector). These proposals had been something very personal to MacSharry.

6.3 EVENT 3: THE COAM MEETING OF 4 FEBRUARY 1991

COM (91) 100 was presented as a 'communication' from the College to the CoAM, not as a formal legislative proposal. MacSharry, in his verbal address, emphasized a more general necessity for a reform of the CAP. This emphasis featured in his presentations to CoAM right through 1991. The two favourite themes were the surpluses situation and the competitive situation of the cereals sector. The stabilizer mechanism was condemned as failing to curb production. Cereals intervention stocks were at 18.5 Mt compared with 11.5 Mt in early 1990. Beef surpluses were at 750,000 t, greater than the previous record set in 1987. Butter surpluses were 260,000 t and the surplus of skimmed milk had reached 335,000 t. The increasing use of grain substitutes had resulted in the demand for cereals as an animal feed falling at a rate of 1.5 Mt to 2.5 Mt per annum.

The removal of the status quo as a policy option was the first tactical objective of the reform team and MacSharry's presentation was part of the achievement of this objective. A mood of 'things can't carry on as they are' was the specific aim of the team. This aim was achieved gradually; event 8, which describes the communiqué from the December Council meeting, shows that not until December did the CoAM agree an explicit, public statement that CAP reform was necessary.

MacSharry's political skills within the Council are discussed as causal links at various points of the CoAM part of the reform process. One of the results of the fieldwork interviews conducted for this work was an emphasis on MacSharry's non-EU political style as affecting the reform process. This was common to different individuals in different institutions involved in the reforms.

The term 'political style' is nebulous, but certain core characteristics of the way MacSharry conducted himself in the CoAM part of the reform process may be identified as relevant. The first set of core characteristics refer to the spin that he put on proposals in his presentations to the CoAM. He had a set of consistent and personal convictions about the CAP and would use these convictions to interpret the reform proposals. There was a definite element of the ownership of the reform ideas; these were *his* proposals for CAP reform. For example, MacSharry was strident in asserting that the ultimate objective of the CAP is the preservation of the maximum number of the 10 million uneconomic farmers in Europe. This is not something that the College would necessarily have concurred with. Indeed, it was an issue in the debate over the agreement of COM (91) 100. This political style meant that the reforms of 1992 came to be associated much more closely with MacSharry (this book adopts the usual description of 'MacSharry reforms') than the reforms of 1984 (milk quotas) or 1988 (stabilizer system) did with the Agriculture Commissioners of the time, Paul Dolsager and Frans Andriessen.

The second set of core characteristics of the MacSharry political style relate to MacSharry's negotiation skills in bilateral or trilateral meetings with individual agriculture ministers and the Council president. Those interviewed as part of the fieldwork for this book all mention MacSharry's experience and background. He had been Irish Agriculture Minister from 1979 to 1981 and Opposition spokesman 1981–1982. Further, he had been involved in various agribusinesses before entering politics. This meant he was well placed to understand the demands of those he was negotiating with in the CoAM. Further, he was able to articulate responses to those demands.

This work does not aim to overplay the importance of the MacSharry style in the reform process. It claims that his way of operating in the CoAM part of the reform process was different from that of previous agriculture commissioners involved in CAP reforms. Participants in, and observers of, the MacSharry reform process note this difference as a causal link which explains part of that reform process.

The identification of specific elements of the reform process affected by political style is difficult. This work judges the influence of MacSharry in terms of specific acts which can be identified as examples of political leadership. Acts of leadership are part of the chain which forms the policy process; they are causal links at various times and at various levels. Acts of political leadership can influence and organize a whole series of individual agent's actions, as is shown throughout this chapter. However, the vaguer claim that MacSharry's political style affected the reform process is much more difficult to establish as a causal factor.

AgraEurope (Green Europe supplement, February 1991) reported a 'small majority' in the CoAM against COM (91) 100. MacSharry reckoned the vote at seven to five against. René Steichen (Luxembourg), then CoAM president, asserted at a press conference that the Council was unanimously behind the need for reform (although there was no agreed statement to that effect) and that more emphasis needs to be placed on 'rural development'. As noted above, the reform team regarded the removal of the status quo as a policy option as the first tactical objective of the CoAM part of the reform process. The rejection of COM (91) 100 did not signal the end of the reform process; it set up the debates in the CoAM over the next 16 months.

Agence Europe charted the cleavages in the Council. The UK, the Netherlands and Denmark issued what one Council secretariat aide called a 'flat refusal' to the direction of COM (91) 100. The southern bloc – Greece, Spain, Portugal and Italy – were in general agreement with the proposals, but their importance was limited by the reform proposals' limited relevance to southern products. Italy spent most of the next 16 months in protracted arguments about its milk quota and was not substantially involved in the evolution of the arguments surrounding the rest of the MacSharry reforms. Germany and Luxembourg requested more concrete statistics before committing themselves. The French position was a microcosm of the wider debate over the MacSharry reforms: Mermaz argued that French farmers could lose FFr 20bn of export subsidies per annum, creating what he termed a 'veritable economic disaster'. However, at the same time, he liked the idea of CAP reform in the terms MacSharry presented it: the concept of the multi-functional farmer, preserving rural areas. His suggestion was that any reform of the CAP in the fundamental way MacSharry had been talking about should be decided by the European Council. The UK position was to support the aim of reform but oppose strongly the idea of modulation.

6.4 EVENT 4: THE 1991/92 PRICE PACKAGE

The agenda of the CoAM after February was occupied by the need for the agreement of the 1991/92 price package. The causal links were the annual CAP price review and the operation of the stabilizer regime. The 1990/91 price package involved some difficult politics in the CoAM. The vagaries of nature determined that the MGQ was not exceeded in the 1991 harvest and thus automatic price cuts for 1991/92 were not triggered.

As described in Chapter 2, the EAGGF budget is sensitive to the world price because of VESs. There is no automatic correlation between the size of surplus production and the EAGGF budget. At the time of the 1991/92 price package, even though the level of cereals production had not exceeded the MGQ, MacSharry claimed that to roll over the 1990/91 prices would lead to the budget breaking the 1988 guideline for EAGGF spending. It has already been established that MacSharry was successful in making the College agree a proposal of price cuts and not the raising of the guideline as a way out of the problem (this was against the wishes of Delors). The CoAM then had to achieve unanimity to raise the spending guideline to prevent tough price cuts. Only the UK's vote prevented this.

The budget problem was resolved by accounting gymnastics. It was an illustration that members of the CoAM care more about price cuts than the budget expenditure of the CAP or meeting the guideline. The stabilizer regime did not impose price cuts automatically in 1991/92 (as it had done in the previous 3 years). This meant that the MacSharry reforms did not become entwined with the annual price package as was the case for 1992/93. The CoAM was not forced to confront the pressing need to reform the CAP, the budget warnings provided by MacSharry not being sufficient. This explains why the reform process was settled in May 1992 and not May 1991.

The battle over the 1991/92 price package and the EAGGF budget occupied the Council's agenda for the first half of 1991. The reform team within the Commission was occupied with this and the restart of the Uruguay Round negotiations. The formal progress of the MacSharry reforms reached a hiatus. It was always intended by the reform team that reform proposals would be brought as a follow-up to COM (91) 100, but during this period the Council's agenda was too crowded for it to start considering CAP reform in detail.

6.5 EVENT 5: AGREEMENT OF COM (91) 258

The College agreed COM (91) 258 (Commission of the European Communities, 1991b) on 9 July 1991. These proposals were effectively the same as those of COM (91) 100; the reference figures of the latter document were inserted as definite figures in COM (91) 258. The two documents have the same name, *The Development and Future of the Common Agricultural Policy*, and are almost identical in their language and numbers. The reason why the reform team did not change COM (91) 100, even though it had been rejected by the CoAM on 4 February, is that COM (91) 100 was deliberately only ever a *communication* to the Council. A small majority against, each member voting for very different reasons, was judged by the reform team as a reasonable point at which to restart the Council part of the reform process.

6.6 EVENT 6: THE OILSEEDS AGREEMENT

The meeting of 23 and 24 September was the first time COM (91) 258 had been considered in the Council of Ministers. However, as after the CoAM meeting in February, the agenda was dominated by another issue. The causal factor was the need to agree an oilseeds support regime by 31 October to meet GATT obligations – specifically, those of a GATT panel ruling. Hence, two factors were involved in delaying the serious consideration of CAP reform by the CoAM in 1991: the 1991/92 price package and the oilseeds dispute.

All members of the CoAM were keen to emphasize that an oilseeds agreement (eventually agreed at the October Council meeting) was in no way a precedent for the proposed reform of the cereals sector. The two issues had been studiously separated on the agenda, hence they could not be discussed at the same time. The international element to the oilseeds dispute prevented this explicit link, and the reform team were unable to use this as a lever to try to agree CAP reform. This reflected similar politics to that involved in the Uruguay Round and described under event 1.

The oilseeds agreement was cited in a number of the fieldwork interviews conducted for this book as 'showing the way' to the reform of the cereals sector. The CoAM agreed an oilseeds regime in October 1991 in which there were DIPs based on local-area cereals yields in order to avoid unbalancing the relationship between rapeseed and sunflower seed production.

The 'showing the way' phenomenon was a causal link because it affected the operation of the CoAM. The institutions paradigm holds that the structure and rules of the CoAM are important variables in the analysis of CAP decision-making. Members of the CoAM represent 12 national agricultural interests. They compete to protect that national agricultural interest in any decision about the CAP. In a period when the level of CAP expenditure has produced political pressure from outside the CoAM, agriculture ministers tend to take defensive attitudes to the protection of their state's agricultural interest.

This defensive attitude means a dislike of change; agricultural ministers fear agreeing to a change in the CAP which gives the appearance of reducing their nation's pay-off from the CAP. 'Appearance' is deliberately used to capture the notion that this is a short-term and political calculation by the ministers; how CAP decisions are immediately received. The existence of a precedent is important in this political calculation. It allows the claim that nothing new has been agreed. Hence the oilseeds compensation scheme sets a precedent for the compensation scheme proposed for the cereals sector by the MacSharry team. The oilseeds agreement shows the way to the agreement of a similar scheme in the cereals sector in May 1992.

The oilseeds regime enacted by the CoAM in October 1991 was rejected by a GATT panel in April 1992, as described in Chapter 4. This has no effect on the precedent referred to above; the central point to the operation of this factor in the reform process was that the CoAM agreed to a compensation scheme based on local-area cereals yield.

The rejection by the GATT panel affected the MacSharry reforms in a more direct way. Proposals for another oilseeds regime were included in the MacSharry reform proposals being considered by the CoAM. The reform proposals considered by the CoAM in May 1992 contained the same type of support regime for the oilseeds and cereals sector. This had never been the case since the EU's oilseeds regime was set up in 1966. The reforms enacted on 22 May maintained the same system of support for both sectors. Thus, GATT obligations and the oilseeds dispute were factors which affected the substance of the MacSharry reforms.

The agreement of the same type of support system for cereals and oilseeds had an effect on the conclusion of the Uruguay Round, as discussed in Chapter 7 on the effects of the MacSharry reforms.

6.7 EVENT 7: PROGRESS OF COM (91) 258 THROUGH THE DUTCH PRESIDENCY

The CoAM of 18 and 19 November 1991 was the first meeting at which there was serious discussion of COM (91) 258. As we shall see, the positions of the member state governments were an elaboration of their February positions.

Throughout the history of the CAP, and in particular in relation to the reforms of 1984 and 1988, the French position had been crucial in the outcome reached by the CoAM. The French farm minister's influence has always tended to be greater than the ten weighted votes that France possesses. The reasons for this influence beyond the strict institutional rule are usually cited as the history of the EU, and the French agriculture interest as the largest beneficiary of CAP support. A more specific factor is that France, since the accession of the UK and Ireland, and the economic trend that has taken the EU from being a net importer to being a net exporter of agricultural commodities, has often been the swing vote in CoAM decisions. This swing position reflects historic contradictions in French government attitudes to the CAP. These contradictions were apparent in the CoAM during the consideration of the MacSharry reform proposals.

Mermaz (French farm minister during the MacSharry reforms) declared a fierce opposition to the proposal of COM (91) 258. He demanded more moderate price cuts, much stronger Community preference, and that the reform be phased in over 5 years. The French attitude to modulation reflected the ambivalence noted above towards the true purpose of the CAP. There was the interest of efficient French farming, France being the EU's largest exporter of agricultural products and the world's second-largest grain exporter. Hence there was the desire to remove the proposal that compensation for land set-aside is not dependent on the size or productivity of a holding. However, the MacSharry cabinet report that the French delegation had an instinctive sympathy with the rhetoric of MacSharry to the effect that a CAP exists for the purpose of keeping farmers on the land and improving farm incomes. The economic reality of the CAP for France is often disguised by the language of its politics and the public justification of the CAP.

The German position was founded on Germany's traditional preoccupation with the income levels of its small farmers, rather than international competitiveness. This had usually taken the form of resisting price cuts; Keichle was not unhappy with MacSharry but wanted more comprehensive compensation for price cuts. As usual, the German farm minister was the most vehement supporter of supply management; Keichle wanted a full 5% cut in milk quotas (with adequate compensation). As described under event 1, the voluntary set-aside scheme of the stabilizer regime was seen by the Germans as a structural measure to avoid price cuts. While it is probably true, as Nedergaard (1994) claims, that Germany 'was from the beginning in favour of the overall content of the 1992 reform', its tactical position at CoAM meetings was more complex than this. In particular, it was a move in the German position that was important in the final settlement of the reform process in the CoAM.

The Belgian position was represented by the veteran farm minister Paul de Keersmaker, who generally followed the German line while insisting on adequate income compensation for cuts in the support price. Luxembourg generally followed the German line through this phase of the reform process.

The UK position, put by Secretary of State John Gummer, was set against any modulation in any regime, but felt most strongly over beef and sheep headage. The basic demands of the British were large price cuts, and any compensation for those cuts to be limited and at a flat rate. The Danish position was similar, emphasizing budget costs as paramount and railing against any discrimination in favour of small farms. Another natural ally in this perspective of the MacSharry reforms, the Netherlands, was hamstrung by occupying the Council presidency in the second half of 1991.

Walsh (Ireland) stated that he would resist any subsidy cut in the beef sector; the fall in grain prices would not benefit the Irish because theirs was a grass-fed livestock industry. Ireland shared with the southern bloc an empathy with regard to the spin MacSharry put on COM (91) 258 of looking after small farmers and having the redistributive aim (as expressed by the '80/20' statistic).

The main agricultural sectors of Spain, Portugal, Greece and Italy were not covered by the MacSharry reforms. They liked the redistributive aims (all have lots of small producers), but were unhappy at yield-based compensation; some of the lowest yields in the Community were in those four countries. The issue of milk quotas was a matter of contention for the southern bloc throughout the reform process – their historic implementation and their current and future levels. This was Italy's main interest in the MacSharry reforms.

The Dutch presidency, during the second half of 1991, was judged by MacSharry and his team as slow in progressing the reform process. The initial blocs defined above remained fixed during the Dutch presidency. This is a typical observation of agricultural policy: if it appears an agreement is not possible within a particular presidency, then less than maximum political effort will be put by the incumbent member state into that project. The rotating presidency of the Council of Ministers is an institutional factor which affects agricultural policy.

During the Dutch presidency CAP reform did, however, occupy part of the agenda of the CoAM, and the reform process maintained momentum. Further, COM (91) 258 was explicitly accepted by the CoAM as the base from which serious negotiations and compromise building could take place.

The December Council meeting agreed that CAP reform was necessary and should be enacted along the following lines: price cuts, compensation, the curtailment of production and respect of the environment. Although these were sufficiently general to be non-controversial, this was the first time the CoAM had moved beyond a nod toward the general principle of CAP reform. The reform team had been confident for some time from informal discussions that this was the case, but saw this public signal as significant; CAP reform would occupy a large share of the agenda of the CoAM in the first 6 months of 1992.

6.8 EVENT 8: THE PORTUGUESE PRESIDENCY

The Portuguese assumed the presidency of the Council in the first 6 months of 1992 (Portugal's first time in this position). The CoAM president, Arlindo Cunha, was an industrious and enthusiastic proponent of CAP reform. He produced five compromise papers before agreement was finally reached on 22 May 1992. The Portuguese presidency was unusual, because although the Council agreed a reform package, little discussion actually took place there, at least in the plenary session. Cunha spent most of his time negotiating with MacSharry and individual member states' representatives. He arranged high-level *ad hoc* meetings of representatives of the national farm ministries between Council sessions, and it was not unusual for him to visit ministers in their respective seats of government.

This political leadership took advantage of the circumstances resulting from the financial reality of the operation of the stabilizer regime. However, this work holds that these circumstances were the necessary and sufficient cause of the CoAM agreeing the reforms on 22 May. The political leadership of Cunha and MacSharry were factors involved in explaining that event. The other causal links are detailed subsequently.

6.9 EVENT 9: THE MARCH 1992 COAM MEETING

It is from the CoAM meeting on 2 and 3 March 1992 that the history of the denouement of the MacSharry reform proposals can be traced. Cunha offered a compromise paper (agreed with MacSharry) to the CoAM; the paper failed to provoke negotiations. The UK, Denmark, the Netherlands,

Belgium and Italy argued that the paper was too detailed and the principles of CAP reform had not yet been established and agreed. Bukman (the Netherlands) railed against a 'salami approach' to CAP reform, in which details were agreed sector by sector instead of the CoAM agreeing a whole package.

The key move was in the German position. Keichle bothered less about the details before principles debate than the fact that the Cunha paper proposed the final target price for cereals at ECU 105 per tonne. The German position was that 130 ECU t^{-1} was the limit to any price cuts. This was the first time the German representative had formally rejected the reform proposals. This encouraged the UK, Denmark, the Netherlands and Belgium formally to reject the Cunha paper.

6.10 EVENT 10: THE 1992/93 PRICE PACKAGE

This lack of progress and direction in the debate over CAP reform prompted MacSharry to threaten draconian price proposals for 1992/93. MacSharry could claim that as a CAP reform was not emerging, he would bring the 1992/93 price proposals under the stabilizer regime. He berated the Council for not taking stock of the reality; MacSharry and his reform team put the following figures into the public domain. Cereal stocks were predicted to reach 25 Mt by the end of 1992 compared with 10 Mt at the end of 1990; milk at 650,000 t and still costing over ECU 5bn; also beef stocks looked to be rising sharply. The next Council meeting was due on 30 March 1992.

It is worth considering an outline of the Cunha paper presented at the Council meeting of 2 and 3 March. Though rejected, this paper describes the development of thinking beyond COM (91) 258. This compromise paper formed the basis of two more that were presented before final agreement was reached. Its salient points are as follows.

The target price at the end of transition for the cereals sector was proposed at ECU 105 t^{-1} compared with ECU 100 t^{-1} in COM (91) 258. The vexed question of beef intervention was tackled: to avoid males of dairy breeds flooding the market a special 'lightweight' intervention regime was proposed. There would be a fixed and falling ceiling for intervention in this category until 1997. The question of headage limits was left as in COM (91) 258.

Finally, there was the introduction of the notion of a base area. This had been something the MacSharry reform team had been working on as a compromise, in response to some member states' fears that there was no effective limit to compensatory payments. The wording of COM (91) 258 implied open-ended compensation; compensatory payments were to be on a per hectare basis with no limit on the number of hectares for which compensatory payments could be claimed.

The base area proposed for the cereals and oilseeds regime sought to define an area for which producers would be entitled to subsidy. Member states could opt for this base area to be specified in terms of individual reference areas or a regional area. With the former, each producer has established an individual base area on the average cultivation from 1989 to 1991. The regional base area is defined by the Commission for each region in the EU on the basis of cultivation over the same 3 years.

If the sum of the individual areas for which subsidy is claimed (basic compensation and set-aside) is greater than the regional base area (whatever the option taken for its calculation), then during the same marketing year, the eligible area per producer will be reduced by the proportion of the overshoot, and the next marketing year will have a special set-aside requirement imposed, uncompensated.

The Commission agreed the 1992/93 price package proposals as a 'rollover' from 1991/92 prices on 18 March. This decision will be shown to have acted as a catalyst in the Council's deliberation over CAP reform. This was a clear act of political leadership by MacSharry and his team, and can be asserted as a factor in the outcome of the reform process.

The analysis presented by MacSharry to that meeting of the College is covered under the operation of the stabilizer regime in Chapter 4. The proposals were not draconian in the sense that MacSharry had threatened at the previous Council meeting, but the 'rollover' did allow MacSharry to make the claim to the CoAM that holding prices at 1991/92 levels within the operation of the stabilizer regime implied an 11% cut in cereals prices.

The price package was presented to the CoAM meeting of 30 and 31 March; it was considered 'quite weak' by *AgraEurope* compared with the threat used by MacSharry at the Council meeting of 2 and 3 March.

The reaction to the price proposals at this meeting was crucial in opening up the prospect of reform by the end of the Portuguese presidency (June 1992). The key reaction was that Mermaz and Keichle both said that they wanted to agree CAP reform before they agreed a price package for 1992/93. The principle that the status quo could not continue, agreed ever since February 1991, had now become political reality.

The cost, to a qualified majority of the CoAM, of agreeing reform was now less than the cost of not doing so. The 1992/93 price package could be ameliorated, incorporated or traded off against CAP reform. The prospect of agreeing a price package which implied an uncompensated 11% cut in cereals prices was the cause of the agreement of the MacSharry reforms. The causal link was the incidence of the political costs of uncompensated price cuts on member state farm ministers.

Nine out of the 12 farm ministers pledged to negotiate a CAP reform package as soon as possible (Belgium, the Netherlands and Denmark requested more discussion before the final horse-trading could begin). There was no new compromise paper from Cunha, who stated that he was not confident enough from bilateral talks to present details, but had a 'feeling' that the Council was moving towards the final negotiations (*AgraEurope*, 3 April 1992). The next meeting was due on 28 April. At this meeting Cunha presented his fourth compromise paper since January.

6.11 EVENT 11: THE AGREEMENT OF COUNCIL REGULATION (EEC) NO. 1765/92 30.6.92, THE 'MACSHARRY REFORMS'

MacSharry rejected the plan Cunha presented to the CoAM meeting of 28 April 1992, so there was no serious discussion of it. The causal factor involved was that a unanimous decision is required by the Council to agree something not proposed by the Commission, whereas a Commission proposal required only a qualified majority to be passed. MacSharry rejected the higher target price proposed for cereals (ECU 112 t^{-1} compared with COM (91) 258 at ECU 100 t^{-1}). This was something which MacSharry was still passionate about in October 1994; he regarded ECU 105 t^{-1} as the limit of EU international competitiveness.

Details of this Cunha compromise paper are patchy. It was similar to the one rejected at the 3 March Council meeting in that the target price was set at ECU 105 t^{-1}. One important difference, however, was the proposal that compensation would be paid on all land set aside, not just the first 7.5 ha as per COM (91) 258. This was part of Gummer's fight against elements of modulation in the MacSharry reform proposals.

The higher eventual target price switched the forecast budget effects: basic compensation payments would fall from ECU 8bn to ECU 6.5bn (there would be a smaller drop in price to compensate), but the set-aside bill would go up from ECU 2.4bn to ECU 3.8bn.

The higher target price also affected the export restitution budget costs. COM (91) 258 implied their virtual elimination by 1996/97 whereas the Cunha compromise and its higher target price implied their maintenance at half of the average current level (an estimated ECU 700m on the budget beyond the end of the transition period). Denmark and the UK wanted larger price cuts, and wanted compensation to be temporary. Gummer liked this paper because, as mentioned above, it removed area limits on the compensation of set-aside. The long-running bane of the UK, namely headage limits in the beef and sheep sectors, remained. Germany, Belgium and Luxembourg thought the price cuts in the compromise were still too severe. The German delegation cited the calculation that a 27% cut in cereals prices meant a 33% fall in German farmers' revenues. Italy, Spain and Greece had a separate agenda in terms of increasing their dairy quotas in return for agreements to respect their implementation. Mermaz thought the paper 'an advance', but took the Commission line that the price cut for the cereals regime was not severe enough.

This set up the two outstanding issues before the marathon May Council meeting (18–22 May) as the level of cereal price cuts and milk quotas.

These would be the core parts of any CAP reform agreement, as they affected all member states. The distinction between the core and the periphery of any package agreed by the CoAM was introduced by Petit *et al.* (1987). The nature of CAP decision-making is the competition of 12 (as the number then stood) national agricultural interests. Policies are not agreed in isolation because of the need to trade off these 12 national agricultural interests; concessions on one part of the CAP beget compromises to different interests on other parts. Policies are agreed together in bundles or 'packages'.

These packages consist of a core which includes policies that affect all or most member states. A core has to be agreed before CAP reform is feasible. After the core is agreed, the special demands of member states are discussed and to some extent satisfied. These demands are characterized by affecting a single member state, or a very small number of member states.

The College met on 1 May and heard evidence from MacSharry that the price of ECU 112 t^{-1} proposed by the presidency would imply a cereals surplus of an extra 8 Mt compared with a price of ECU 105 t^{-1}. The College 'disapproved' (*Agence Europe*, 1 May 1992, no. 5721) of the fact that the CAP reform debate seemed to have descended into bargaining over the level of price reductions.

A further Cunha paper was submitted to the May meeting, and this time provided the basis for agreement. The 1992/93 price package was part of the agenda as well. MacSharry hinted that the Commission could make certain elements of the price package 'more flexible' – that is, less tight – in return for agreement on CAP reform; for example, a reduction to 3.5% (instead of the 5% proposed) in the CRL and a reduction in supplementary taxes were both mentioned. This is how MacSharry and his reform team used the need to agree 1992/93 prices to lever the CAP reform.

The debate centred on cereal price reduction (and compensation) and milk quotas. MacSharry and his team persuaded the College to reject the 24% reduction in cereal support suggested by the presidency; they argued that this level of price cuts would require 22% set-aside to keep surpluses stable. MacSharry wanted the College to propose 3% reductions in the overall milk quota of the EU, but the requests of Italy, Spain and Greece for their national quota to be raised were such that overall Community production would have been increased by 3%. These requests had been rejected by other members of the CoAM as 'unjustified'.

The week 18–22 May was characterized by trilateral meetings of MacSharry, Cunha and individual ministers. This is standard CoAM practice (which marks it out from other Councils in the way that it operates); the plenary sessions lasted for about 4 h during the week.

MacSharry considered 29% to be the minimum required reduction in cereals prices, which compares with 24% in the most recent Cunha paper and the 35% of the Commission proposal. By the morning of 20 May, the presidency's fifth document on CAP reform (in the 5 months since Portugal had taken over) was emerging. Cereal prices were to be reduced by 29%, to a target price of ECU 110 t^{-1} and an intervention price of ECU 100 t^{-1}. In the dairy sector, national milk quotas would be 'adjusted' in Italy, Spain and Greece on the condition that previous quotas had been applied in a manner satisfactory to the Council.

The 'northern' bloc in the CoAM (Belgium, the UK, Luxembourg, the Netherlands, and Denmark) balked at any concessions on milk quotas. They complained of special treatment for countries which had not respected the system (Spain, Greece and Portugal were exonerated on account of their recent accession to the Commission). Italy was singled out as the main transgressor; it had not implemented the 1984 milk quota system and yet it wanted its quota increased. The Italian Minister of Agriculture, Goria, had been granted the weapon of Luxembourg compromise (from Giulio Andreotti) to veto the entire package.

This situation was resolved with the Italian quota issue separated off and dealt with over subsequent months. The dairy sector itself represents probably the greatest dilution of the original MacSharry proposals.

The College agreed to the 20 May document on the recommendation of MacSharry that it enjoyed sufficient support in the CoAM to bring everyone else in 'in its wake'. This document was the core of the outcome of the MacSharry reform process (the level of cereals prices and related compensation).

Thursday 21 May was occupied by the agreement of the periphery. Various issues and special member states' demands were linked to produce final agreement on Friday 22 May. The UK gained some concessions in the sheepmeat sector. German demands were ameliorated by the agreement of a 1992/93 price package (at the same time as MacSharry) less strict than the rollover of 1991/92 prices under the stabilizer regime. The Irish gained concessions with respect to the slaughter of male cattle. The French secured some details about fodder grasses and ensilage corn.

The main compromises made in the week 18–22 May are as follows. The reduction in support prices was set at 29% over 3 years. Each individual producer was to be fully compensated through direct payments. To be eligible for these compensatory payments there was an initial set-aside requirement of 15% – for which further compensation was available through set-aside payments. Neither of these payments was modulated except in the sense that small producers below a defined limit could be exempted from the set-aside requirements.

In the beef sector support prices would be cut by 15% over 3 years. There were some new limits on beef intervention, but the 'safety net system' continued. Other measures were agreed as a concession to the Irish because their beef is grass-fed and would not benefit from a fall in animal feed prices (for example, the introduction of a winter beef premium, otherwise known as 'de-seasonalization'). The headage limits for premium payments remained for beef, but were removed for suckler cows.

The sheepmeat represented the substance of the concession to the UK. Half of the premium would be paid beyond the headage limit and 1989 would be added to 1990 and 1991 as possible reference years.

The proposals of MacSharry for the dairy sector were very diluted in the final agreement. The intervention price for butter was reduced by 5% (rather than the 15% proposed) and for skimmed milk powder there was no change in price. The original DG VI document argued for a 5% cut in the global milk quota; in fact there was no agreement to cut the quota.

The 1992/93 price agreement was the central focus of the German farm minister's negotiating demands. In the cereals sector, the basic CRL was reduced to zero (from 5% in 1991/92). The intervention price was cut by 3% according to the stabilizer mechanism, but the supplementary CRL of 3% was dropped. This meant that there was an effective price rise of 2% compared with MacSharry's oft-quoted threat of a cut of 11%.

6.12 CONCLUSIONS

This chapter describes a chain of events which together form the policy process that led to the enactment of the MacSharry reforms. Each section in the chain and the causal links between the sections of the chain are looked at. The policy process is element 2 of the CAP decision-making system as set out in Chapter 1. Chapter 4 describes the possible causes of the MacSharry reform process; this formed element 1 of the CAP decision-making system (Chapter 1). Element 3, the output of the CAP decision-making system, is covered in the next chapter on the effects of the MacSharry reforms.

The policy process associated with the MacSharry reforms has been described through the institutions framework, a choice of framework that is justified in Chapter 5. Chapter 8 checks whether the policy process detailed in this chapter can be better interpreted using one of the alternative frameworks outlined in Chapter 5. Chapter 5 described 'better' in this context as meaning the provision of a more coherent, clear and concise interpretation of the policy process associated with a CAP reform.

The argument of this chapter can be summarized using the three questions posed for the MacSharry reform policy process in Chapter 1 of this book: why were the MacSharry reforms enacted, when were they enacted and why were they enacted in the way that they were? These questions help to provide an understanding of the key elements in this complex process.

As emphasized throughout this book, CAP reforms should be understood as the outcome of a process. The factors involved in the initiating of a reform process are not necessarily the same factors as those that conclude the process. The MacSharry reform process started as the result of the domestic pressure for CAP reform (the operation of the stabilizer mechanism) and international pressure in the Uruguay Round interacting with an institutional structure in which the Agriculture Commissioner had responsibility for the reactions to both these pressures. Commissioner MacSharry's distinctive political personality has been emphasized in this chapter and the issue of the role of individuals in the public choice paradigm is considered in Section 9.4.1. However, this chapter holds that the interaction was structured and strongly affected by the institutional context in which it took place.

The institutional structure of CAP decision-making translated the pressures noted in Chapter 4 into two driving forces for CAP reform. The first was through an Agriculture Commissioner and a President with an ambition for reform and receptive to ideas being produced by the special studies unit of DG VI. The second, after the collapse of the Uruguay Round at Brussels, was the demand for a more flexible negotiating stance.

The MacSharry reform process was concluded when the stabilizer mechanism implied nominal price cuts for 1992/3 that were too swingeing for the collective political stomach of the CoAM. There was a lower political cost attached to enacting a reform than to preserving the status quo. The form of the MacSharry reforms was determined by the concurrence of the CoAM's agenda of minimizing the political costs of any reform by compensating price cuts and the Commission's ambitions for a CAP reform which controlled production and provided a credible negotiating stance with the USA.

Chapter 7

The Effects of the MacSharry Reforms

7.1 INTRODUCTION

It is important to establish initially what this chapter does not intend to cover. It is not using a description of the effects of the MacSharry reforms as an explanation of why those reforms happened, when they happened and the way in which they happened; it is not part of a functional explanation of the MacSharry reforms. Rather, it describes the effects of the MacSharry reforms in a way that attempts to avoid *ex post* explanation and justification of those reforms. This serves two functions. First, it gives a completeness to the account of the MacSharry reforms. Second, and more importantly, it helps emphasize the claim of the institutions paradigm that CAP reforms under the present institutional arrangement of decision-making will never affect the underlying stability of the CAP. Chapter 1 defines a radical reform as one which affects the distribution at national level of the costs and benefits of the CAP. The MacSharry reforms are shown in this chapter to be something less than radical (in the sense discussed in Section 1.5).

This claim exists alongside the observation that a switch to quotas (1984), tying price changes to output (1988) and introducing DIPs (1992) are significant changes in how the support of the CAP was delivered. The MacSharry reforms are in other senses radical: substantial cuts in nominal support prices; a shift in the burden of support from consumers to taxpayers; the introduction of direct income compensatory payments linked to participation in a set-aside scheme. The definition of 'radical' introduced in Chapter 1 was linked to the idea of CAP decisions, including CAP reforms, as the product of a decision-making system. The configuration of this decision-making system constrains the potential products of that system. The CoAM, at the time of the MacSharry reforms, was a competition of 12 national agricultural interests and CAP reform is the result of this competition. The judgement of whether a particular reform is radical or not should be made in terms of those 12. The final section of this chapter looks at the literature on proxies to measure national agricultural interests.

The effects of the MacSharry reforms can be described in two dimensions. The first is the measured economic and financial effects in the operation of the CAP since 1992. This includes, for example, the budget effect; that is, the effect on production and the level and distribution of farm incomes. The second dimension follows from an understanding of the MacSharry reforms as the result of processes produced by a set of circumstances. These circumstances were described in Chapter 5; the CAP was subject to international negotiations in the Uruguay Round, the oilseeds dispute and the operation of the stabilizer regime, which was producing politically unacceptable price cuts. These affected the MacSharry reforms, and the consequences of the reforms for these circumstances is the second dimension considered in this chapter.

7.2 THE ECONOMIC AND FINANCIAL EFFECTS OF THE MACSHARRY REFORMS

The strict economic and financial consequences of the MacSharry reforms of 1992 can be judged over the short or medium term. This section surveys the evidence for both and describes the predictions produced by economic models of the CAP for the 10 years after the MacSharry reforms. These are *ex ante* analyses of how the reforms are expected to affect the direction of the CAP. These *ex ante* analyses cover a time period from 1 to 10 years after the implementation of the reforms. If the reforms mark a significant change in the operations and effects of the CAP, this will be picked up by agricultural economists. The short-term economic and financial effects can be described by observed economic history: the published figures of how the CAP has operated since 1992. This is *ex post* analysis. This book is written in 1997 as the transition period of the MacSharry reforms is ending and the CAP is operating under the final levels of support prices and compensatory payments. The amount of evidence by which to judge the effects of the MacSharry reforms is therefore limited, but some tentative conclusions can be reached.

7.2.1 A Medium-term *Ex Ante* Perspective on the Economic and Financial Effects of the MacSharry Reforms

An expert panel of independent agricultural economists used a series of models to analyse the consequences of the MacSharry reforms for the 10 years following 1992 (*European Economy* no. 4, 1994, 'EU agricultural policy for the 21st century'). These models cover international agricultural markets as well as EU domestic markets. A number of different scenarios are considered; of interest to this chapter is the comparison of the MacSharry reforms with the option of having not reformed the CAP in 1992. The latter option assumes that the stabilizer regime would have continued to operate starting with 1991/92 institutional prices in 1992/93. A number of different aspects of the effects of these two options were considered; the results for the EU budget, EU production levels and EU farm incomes over the time period 1992–2001 are presented here. These aspects are highlighted in this chapter because they are directly mentioned in the foreword to COM (91) 100 (Commission of the European Communities, 1991a).

The results of the economic models of the panel of independent experts presented in this section are juxtaposed with figures for what has actually happened in the 3 transition years of the MacSharry reforms presented in Section 2.2. Scenario 1 (Table 7.1) is the no-reform option and scenario 2 (Table 7.2) represents the MacSharry reforms of 1992.

Scenario 2 necessarily contained predictions of what will happen after the end of the transition period of the MacSharry reforms in 1996. The

Table 7.1. Scenario 1: the no-reform option (figures in million tonnes). (*Source*: *European Economy*, no. 4, 1994.)

Production	1992	2001	% annual change 1992–2001
Grains	163.9	194.5	1.9
Oilseeds	n/a	n/a	n/a
Oilcakes	14.9	18.9	2.7
Oils	3.9	n/a	n/a
Other grain substitutes	21.5	n/a	n/a
Beef	8.1	8.2	0.1
Pork and poultry	25.6	31.9	2.5
Milk	96.7	96.7	0
Sugar	15.9	15.9	0

Table 7.2. Scenario 2: predicted effects of the MacSharry reform. (*Source*: *European Economy*, no. 4, 1994.)

Production	1992	2001	% annual change 1992–2001
Grains	163.9	158.4	−0.3
Oilseeds	n/a	n/a	n/a
Oilcakes	14.9	16.1	0.8
Oils	n/a	n/a	n/a
Other grain substitutes	21.5	n/a	n/a
Beef	8.1	7.9	−0.2
Pork and poultry	25.6	34.0	3.3
Milk	96.7	94.7	−0.2
Sugar	15.9	15.9	0

main assumption is that EU domestic prices will fall between 1996 and 2001 at roughly the same rate as they had done during the 1980s, but that world market prices will develop at a 'more favourable rate' (i.e. increase, or decrease).

The most interesting result is that for the grains sector. The MacSharry reforms are predicted to reduce the level of production in that sector, compared to a 1.9% growth in production if the CAP had carried on unreformed over the period.

7.2.2 Budget Effects

For scenario 1, the model of the international agricultural markets used in the *European Economy* report provides the result that world market prices would have fallen marginally through the 1990s, while EU domestic prices would have fallen more substantially. This result means that in scenario 1 the budget costs of export subsidies would have fallen, even though production increased (see Section 7.2.1). Overall budgetary transfers would have fallen by ECU 1.5bn from 1992 to 2001. However, storage costs would have been expected to increase, reflecting a higher level of publicly held intervention stocks. This leaves predicted total budget costs increasing by about 0.3% per annum from 1992 to 2001.

In scenario 2, the introduction of DIPs will increase the level of budgetary transfer payments. Exports and the level of export subsidies decrease but this is outweighed by the increase in direct payments to the level where total budget transfers are estimated to increase by ECU 4.2bn. This gives an increase in the total EAGGF guarantee expenditure budget of 2.3% per annum for the years 1992–2001.

Ackrill *et al.* (1994), using a different model focusing specifically on the EU cereals sector, produce a similar kind of result: the total budget costs of MacSharry compared to a continuation of the stabilizer regime are greater. However, these expenditures are more stable and predictable over the medium term than previous EAGGF spending; that is, set-aside payments and compensatory payments will increasingly dominate EAGGF expenditures, and these are less sensitive to fluctuations in the world price than export refund payments, which have in the past upset the reliability of EU budget forecasts.

7.2.3 Farm Incomes

The results of the models follow the predictions of economic theory (set out in Chapter 2) that a switch from a system of market price support to DIPs is a more efficient way of supporting farm incomes. Scenario 2 generates an improvement of ECU 5.7bn in aggregate real farm incomes compared with scenario 1.

Nardone and Lopez (1994) estimate the economic welfare effects of the MacSharry reforms in the EU wheat sector in the first year of the

Table 7.3. Computed welfare estimates of the MacSharry reforms 1992/93–1993/94. (*Source*: Nardone and Lopez 1994.)

	1992	1993	Change
Farm price	155	130	–25
Area planted	16.81	15.46	–1.35
Production	90.13	80.84	–9.29
Consumption	64.97	68.13	+3.16
Exports	25.16	12.71	–12.45
Welfare effects:			
Producer surplus gain	5860.10	5841.18	–18.92
Consumer surplus loss	5283.17	3620.82	+1662.36
Budget expenditures from:	1886.85	2888.63	–1001.78
Export restitutions	1886.85	635.35	+1251.50
Direct subsidies	0	2253.28	–2253.28
Deadweight loss	1309.93	668.27	–641.66
(% of producer gain)	(22.35)	(11.44)	

Note: Farm price figure is in ECU t^{-1}; area planted in million hectares; production, consumption and exports in 1000 tonnes; and policy transfers are all in million ECU.

transition period of the reforms. Table 7.3 describes their results. The main points are that EU wheat farmers suffered a relatively small welfare loss in the first year of the MacSharry reforms (in contrast with the result presented in Section 2.1.3). Consumers gained and EU budget expenditures rose because of direct income and set-aside compensation payments. This last effect was partially offset by the reduced level of export restitutions because of the lower gap between the EU internal market price and the world price.

Nardone and Lopez (1994, p. 388) conclude that 'EU wheat producers are not expected to gain from the new CAP regime and that the regime will exacerbate the EU budget'. On their calculations, the level of set-aside compensation is insufficient to cover the lost income on the production that could have taken place on land set aside.

7.2.4 A Significant Change in Direction?

The medium-term *ex ante* scenario for the post-MacSharry CAP is less production, higher EAGGF budget costs and higher farm incomes (the Nardone and Lopez result is for the first year only) than would have otherwise been the case. The quantitative extent of these changes is provided in the preceding sections describing the results of *European Economy* (no. 4, 1994). This work, combined with the results of Nardone and Lopez (1994), supports the tentative conclusion that, *ex ante*, the MacSharry reforms do not represent a significant change in the direction and consequences of the CAP. Despite the capping of compensatory payments, the capacity of the agricultural budget to grow in the medium term still remains.

7.3 *EX POST* ANALYSIS: THE SHORT-TERM ECONOMIC AND FINANCIAL CONSEQUENCES OF THE MACSHARRY REFORMS

This section presents evidence of the effects of the MacSharry reforms in the following fields: the EU cereals market, concentrating on production levels and the amount of publicly held stocks; the budget, both its pattern and its overall level; and, briefly, farm incomes.

7.3.1 The Cereals Market

The transition period of the MacSharry reforms started in the marketing year 1993/94, so their effect is first shown in the figures for 1994 in Table 7.4. The two notable features are the containment of production despite the continued trend of the growth in yields, and the decline in the ending stocks figure in the years 1994 and 1995.

Production was the same in 1993 as 1995, despite a growth in yields, because the area cultivated declined. This decline may be attributed to the set-aside requirements of the MacSharry reforms (although small farms are exempt). The increase in yields resulted partly from the overall progress of technology but also from the phenomenon of *slippage*. Within an individual farm, farmland is not of uniform quality. The requirement to set aside 15% of land in order to participate in the compensatory payments scheme will be fulfilled by the least productive 15% of a farmer's land. Alternatively, for rotational set-aside, land out of production for a year may often become more productive when recultivated. Either way, total farm production will fall by less than 15% and the farm's average yield (on land cultivated) will rise.

As mentioned in Chapter 2, surpluses are disposed of in two ways: through publicly held intervention stocks (or ending stocks), or by directly

Table 7.4. EU12 cereals supply balance. (*Source*: USDA.)

	1989	1990	1991	1992	1993	1994	1995
Area (Mha)	37.4	36.1	36.2	35.3	32.1	32.0	32.0
Yield (t ha^{-1})	4.62	4.72	5.02	4.78	5.09	5.01	5.11
Production	173	170	182	169	164	160	164
Consumption	151	142	145	139	151	153	155
of which feed use	89	82	83	77	88	90	93
Exports	57	55	58	60	57	52	49
Imports	32	32	33	31	34	35	37
Ending stocks	25	31	42	43	32	23	19

Note: All figures in million tonnes unless stated.

subsidizing their export. The decision of how the Commission manages production levels depends on the gap between the EU price and the world price. The EU price has been reduced through MacSharry, but in addition to set-aside, a further significant factor in the decline of publicly held intervention stocks has been the increase in the world price. This reduction in the gap between the EU domestic price and the world price has affected the storage costs and export subsidy items of the EAGGF budget.

7.3.2 Budget Effects

The circumstances of the EU budget and the CAP's effect on those circumstances have been widely cited as the factors behind previous reforms of the CAP (see Chapter 3). The budget effects of MacSharry would intuitively appear to be adverse; for a given level of protection, the switch from price support (consumers' burden) to DIPs (taxpayers' burden) would seem to impose a greater share of the cost of the CAP on the EU budget. COM (91) 100 (Commission of the European Communities, 1991a) mentions the growth in surpluses as a problem and contrasts the growth in EAGGF spending and farmers' incomes. However, the failure to contain the rate of growth of EAGGF spending is not criticized *per se*. MacSharry identified surpluses

Table 7.5. Summary figures for total EAGGF expenditure (billion ECU). (*Source*: European Commission (*Agricultural Situation in the Community, Official Journal*.)

	1990	1993	1994	1995	1996
Total	26.53	34.43	32.21	36.89	40.82
Cereals	7.87	10.68	12.70	14.57	17.19
Dairy	4.97	5.26	4.24	4.26	4.21
Beef and veal	2.83	3.99	3.46	4.88	5.46

Note: The figures for 1996 are forecasts.

Table 7.6. Budget of the CAP cereals sector, 1993–1996 (billion ECU). (*Source*: European Commission (*Agricultural Situation in the Community, Official Journal*).)

	1993	1994	1995	1996
Total	10.68	12.70	14.57	17.19
Export refunds	2.79	1.73	1.26	0.91
Cost of storage	2.72	0.28	0.12	0.28
Compensatory payments	3.29	9.01	10.86	13.47
Set-aside	0	1.28	2.39	2.08
Other	n/a	n/a	n/a	n/a

Note: The figures for 1996 are forecasts.

as the issue and warned that his reforms should not be interpreted as a reduction in agricultural support spending. Indeed, COM (91) 258 (Commission of the European Communities, 1991b) estimated that, overall, the annual budget would be ECU 2.3bn higher if the reforms proposed in that document were enacted. It was also admitted that this estimate was sensitive to parameters chosen and longer-term market trends. This was the background to the MacSharry reforms; it was not enacted in a budget crisis in the way that the reforms of 1984 and 1988 were.

The most notable feature of Table 7.5 is the growth in the size of the cereals sector budget; between 1993 and 1994 it grew by 18.9%, in 1995 by 14.7% and is expected to increase by 18.0% in 1996. The fact that the total EAGGF budget for 1994 was within its guideline is not due to a constraint on the growth of the cereals sector.

Table 7.6 gives a breakdown of the cereals sector budget from 1993 to 1996. The salient features are the reduction of the storage costs to almost zero, the reduction of the level of export subsidies, and the growth in SAPs and CPs. The first two of these effects, on storage costs and export subsidies, occurred during the transition period of the MacSharry reforms; however, these figures have been much more substantially affected by a favourable trend in world prices. Chapter 2 describes how favourable trends in world prices can lead to the Commission reducing the level of publicly held intervention stocks.

The growth in the level of compensatory payments (CPs) and set-aside payments (SAPs) can be much more directly attributed to the MacSharry reforms. These two payments accounted for 30% of the cereals sector budget in 1993 and are due to be 90% in 1996.

7.3.3 Farm Incomes

The need to improve farm incomes was specifically mentioned in the foreword of COM (91) 100 as an objective of the reform proposals. The following evidence on the level of farm incomes is from *Agricultural Situation in the Community* (1995).

The 1993/94 accounting year is where the first observations appear of the effect of the MacSharry reforms on farm incomes. Across all farm types and member states average farm incomes fell by 6% in real terms relative to 1992/93. However, income increases were observed for sectors most directly affected by the MacSharry reforms; for example, the average income of arable farms rose by 10%, that of drystock farms rose by 4%, and the average income of dairy farms increased by 6%, whereas the average income of pigs and/or poultry farms dropped substantially, by 15% of their 1992/93 income level.

Such complete evidence has not been published for 1994/95. *Agricultural Situation in the Community* (1995) suggests that preliminary evidence for 1995 shows that incomes in the cereals sector have continued to improve, and also that incomes in pigmeat farming have begun to recover. However, there are negative income trends for most other livestock types, especially poultry, sheep and cattle farming.

The 80/20 statistic which played such a psychological role in the reform process was never statistically supported by the Commission, and no evidence has been provided as to how the MacSharry reforms have affected it. However, the kind of disparity that underlies the 80/20 figure has not disappeared. In 1993/4 the top 20% of EU farms had an average income as high as ECU 42,000 and the bottom 20% as low as ECU 4000.

It should be noted that high farm incomes in a particular agricultural sector are generally a function of favourable market conditions and technical progress. Market conditions can be affected by a whole range of economic factors as well as by CAP reforms. It is therefore often difficult to accurately attribute the trends in farm incomes to changes in the support regime, especially over such a short time period as 3 years. However, it is still interesting to note that there is not even weak evidence that farm incomes have been damaged by the reforms.

7.4 THE EFFECTS OF MACSHARRY AS SUPPORT FOR THE INSTITUTIONS FRAMEWORK OF CAP DECISION-MAKING

7.4.1 The Budget and Trade Effect as a Measure of Each Member State's Net Pay-off from the CAP

The operation of the CAP has redistributive effects between member states. There are two principal effects: on budget and on trade. The EAGGF budget effect can be calculated by looking at the geographical pattern of CAP expenditure and the burden of the EAGGF distributed between member states. In a similar way, the operation of the CAP creates a preferential trade effect within the EU. Members can export commodities to partner countries at prices above world trading levels, thereby gaining revenues, while the importing member state has to pay more for the commodities than if that state traded at world price levels. This redistributive effect, like the budget effect of the CAP, can be calculated at member state level. These two effects together are known as the budget and trade effect (BTE). The BTE has been used by economists over a long period to examine the distributional effect of the CAP between member states (Morris, 1980; Buckwell *et al.* 1982; Ardy, 1988; Brown, 1988; Ackrill *et al.,* 1995).

The measure of BTE is expressed in Table 7.7 as a percentage of that country's GDP, thereby giving an insight into the relative size of the costs and benefits of the CAP to each member state.

7.4.2 Stability in the Distribution of the BTE between Member States, 1981–1992

The results of various studies of the distributional effects of the CAP over time are presented in Table 7.7. The salient feature is the stability in each

Table 7.7. Budget and trade effort (BTE) as a percentage of GDP, 1981–1993, for EU member states. (*Source*: B = Brown (1988); N = Ackrill *et al*. (1995); M = MAFF (1995).)

Year	Study	Belg.,Lux.	DK	D	GR	SP	FR	IRE	IT	NL	PT	UK
1981	B	–0.1	1.1	–0.3	2.1	—	0.3	4.2	0	0.7	—	–0.5
1982	B	–0.1	1.4	–0.3	2.0	—	0.3	4.6	–0.1	0.7	—	–0.6
1983	B	–0.1	1.4	–0.3	1.7	—	0.1	5.2	0.1	1.0	—	–0.5
1984	B	–0.1	1.3	–0.3	2.0	—	0.2	4.9	0	0.9	—	–0.5
1985	B	–0.1	1.2	–0.4	2.2	—	0.4	5.0	–0.1	0.6	—	–0.4
1986	—	—	—	—	—	—	—	—	—	—	—	—
1987	N	–0.3	1.4	–0.2	2.0	–0.3	0.4	6.4	–0.4	1.1	–0.6	–0.5
1988	N	–0.3	1.3	–0.4	2.4	–0.1	0.3	5.9	–0.3	1.2	–0.2	–0.4
1989	N	–0.1	1.2	–0.3	2.5	–0.1	0.2	5.1	–0.3	1.4	–0.1	–0.4
1990	N	–0.1	1.3	–0.3	2.5	–0.1	0.3	5.8	–0.3	1.0	0	–0.4
1991	N	–0.1	1.3	–0.3	2.5	0.1	0.3	6.0	–0.3	0.6	–0.3	–0.2
1992	N	–0.1	1.2	–0.4	2.5	0.2	0.3	5.7	–0.3	0.5	–0.1	–0.3
1993	M	0	0.8	–0.3	2.8	0.3	0.3	4.4	–0.3	0.3	–0.5	–0.2

country's net pay-off from the CAP. This fits in with the discussion of stability in the CAP in Chapter 3.

Shifts in BTE positions can sometimes be accounted for by the fact that different studies have employed slightly different methodologies for calculation. However, the significant changes in net positions since 1980 can be summarized as follows. Spain and Portugal, in the period since accession, have gone from having a negative net pay-off from the CAP to being positive beneficiaries (although in the case of Portugal it took the MacSharry reforms to achieve this position – see below). Italy went from being a net beneficiary in the early 1980s to being a net loser. The net pay-off of the Netherlands from the CAP under the stabilizer regime declined significantly. Although the aggregate figures presented in Table 7.7 do not show it, this decline was almost entirely due to a deteriorating budget effect. Between 1989 and 1991, the positive budget effect for the Netherlands from the CAP declined by 88%.

7.4.3 The Effect of the MacSharry Reforms on the National Distribution of the Budget and Trade Effect (BTE) of the CAP

Table 7.8 presents the results from the study of Ackrill *et al.* (1995), which examined the distributional effects of the MacSharry reforms between member states. The pre-MacSharry figures are those calculated for 1992. The results for the effects of the MacSharry reforms are obtained by supposing that the reforms had come into effect immediately in 1992; that is, that there had been no transition period between 1993 and 1996, so that the price levels agreed for 1995/96 were used.

The central feature is the stability in most countries' net position after

Table 7.8. Effects of the MacSharry reforms on the budget and trade effort (BTE) for EU member state. (*Source*: Ackrill *et al.*, 1995.)

	BTE pre-MacSharry	BTE post-MacSharry	Change	Total BTE as % GDP pre-MacSharry	Total BTE as % GDP post-MacSharry
Belgium, Luxembourg	–270	–256	14	0.1	0.1
Denmark	1341	1476	135	1.2	1.4
Germany	–6495	–6416	79	0.4	0.4
Greece	1880	2006	125	2.5	2.7
Spain	1157	1400	243	0.2	0.3
France	3506	3130	–376	0.3	0.3
Ireland	2229	1978	–251	5.7	5.0
Italy	–2384	–2327	58	0.3	0.3
The Netherlands	1646	1376	–270	0.6	0.5
Portugal	–103	98	201	0.1	0.3
UK	–2808	–2650	158	0.3	0.3

Note: The GDP figures are rounded to one decimal place.

the MacSharry reforms. The conclusion of the authors is that the overall distribution at a member state level of the costs and benefits of the CAP has not changed markedly as a result of the MacSharry reforms: 'we see the MacSharry reforms not having a significant impact on the overall position of most countries' (Ackrill *et al.*, 1995, p. 14).

Within this broad conclusion, a number of changes can be noted. The decline in the net position of the Netherlands has been slowed, providing perhaps some limited evidence that members of the CoAM do pay attention to such measures of a national pay-off from the CAP when they are deciding their negotiating stance. As mentioned above, the net position of the Portuguese has become positive for the first time since accession. Finally, and perhaps ironically, the Irish seem the most notable losers from the MacSharry reforms. It should be noted that there are aspects in the methodology of calculating the BTE which may account for some of these changes.

7.4.4 Evidence of Stability in the CAP

Tables 7.7 and 7.8 show the distributional effects of the CAP over the period of the three most notable reforms of the policy: the introduction of milk quotas in 1984; the stabilizer regime in 1988; and the MacSharry reforms of 1992. Since 1980 the net BTE positions of most member states have exhibited a high degree of stability, in both their direction and their size, despite these three major reforms.

The effects of the MacSharry reforms in this respect have been described in this chapter to provide alternative support for one of the main themes running through this work: that CAP reforms, including MacSharry, have always fallen short of what might be considered radical. The BTE measure helps provide alternative support for the view of the CAP decision-making system which sees reform as turbulence within what essentially is long-run stability. The institutions paradigm of the CAP, outlined in Chapter 3, was designed to explain this long-run stability in each country's net pay-off from the CAP.

7.5 THE URUGUAY ROUND

Chapter 5 describes the chronology of the Uruguay Round of GATT negotiations from the Punta del Este agreement of September 1986 to the enactment of the MacSharry reforms. Chapter 6 gives details of the concurrence of the timetable for the round and the timetable of the MacSharry reforms between 1990 and 1992. It explores the demands of the international arena as a possible causal factor in domestic agricultural policy changes. This section completes the story by considering the effect of the enactment of the MacSharry reforms on the final agreement. Similarly, Section 7.6 finishes the account of the oilseeds dispute.

7.5.1 The Conclusion of the Uruguay Round 1992–1993

In some respects the MacSharry reforms of May 1992 went beyond the requirements of the Draft Final Act, described in Chapter 6. Briefly, the Commission offered the argument around May and June 1992 that the reforms would eliminate the need for most export subsidies by 1996/97. The Draft Final Act offered the reduction of export subsidies over 6 years as the basis for negotiation – hence the expectation at the time of the enactment of the CAP reforms that the context of negotiations would be changed. To an extent, that proved to be true, but the differences between the USA and EU in the Uruguay Round proved durable during the next 5 months.

The EU was 'understood' (FT, 12 June 1992) to be offering only a 20% reduction in the volume of export subsidies. The USA was insisting on the Draft Final Act figure of 24%. The USA demanded that the 15% set-aside for 1992/93 should be 'set in concrete' for future years. The classification of DIPs was an issue. The EU contended that DIPs linked to output restraint (in this case set-aside requirements) should be indefinitely exempt from GATT reductions; this 'philosophy' should be agreed to continue into future rounds. Finally, the EU wanted a 'peace clause' binding the USA to use the GATT dispute resolution system ahead of the unilateral 'super 301' sanction.

The Uruguay Round drifted through late summer 1992 on this footing. However, the talks broke down on 21 October, with the USA introducing the threat of sanctions in the accompanying oilseeds dispute. EU products to the value of $300bn were targeted for punitive tariffs if the oilseeds dispute was not settled by 5 December. This separate dispute (about the violation of existing GATT rules, i.e. nothing to do with the Uruguay Round) became pivotal in the final negotiations on agriculture. This linking of the two issues was a clear USA tactic.

An oilseeds production limitation was subsequently added to the list of outstanding disputes between the EU and USA in the Uruguay Round. The publicity during October 1992 was of the two sides being separated by 500,000 t of soybean. The USA was insisting on cutting EU output to 9 Mt. The EU claimed that limits below 9.5 Mt would involve tampering with the agreement on CAP reform of May 1992. The 500,000 t was not important; what was important was that the USA, by pushing the oilseeds dispute, had handed the French negotiators a symbol or banner under which to organize their opposition to any major agricultural trade agreement. That banner was the 'compatibility' argument. This compatibility demand was the source of Delors's failure to be seen to be unequivocally backing the Andriessen–MacSharry line in the Uruguay Round negotiations. This issue was cited as a reason for the brief MacSharry resignation after the Chicago meeting with Madigan on 1 and 2 November.

7.5.2 Blair House Accord

The Blair House Accord, reached at Blair House, near Washington, DC, on 20 November 1992, averted the USA threat of a trade war beginning on 5 December. The following areas were agreed. On export subsidies, beginning in 1994 the EU would reduce the volume of its subsidized exports by 21% (compared with the Draft Final Act figure of 24%) over 6 years. The value of its subsidized exports would be cut by 36% on the same time scale. Internal subsidies would be cut by 20%. The USA accepted the EU position that DIPs linked to set-aside should not be treated as trade-distorting. The oilseeds dispute was resolved by the EU agreeing to reduce the amount of land used to cultivate oilseeds from 13.5 Mha to 11 Mha, starting with an initial reduction of 15% of that area and continuing with at least a 10% reduction in the ensuing years. In addition, EU farmers would be able to produce up to 1 Mt of oilseeds a year for industrial uses, such as biofuel production. The USA, in return for the EU conforming to these requirements, agreed to a peace clause in which both sides agreed not to demand further GATT investigations into the domestic support regimes so long as both sides respected the commitments made in the Uruguay Round. This provided the substance for the claim by MacSharry that the Blair House Accord as part of a Uruguay Round agreed would put the CAP on a legal footing within GATT for the first time.

The rebalancing issue was resolved by a USA pledge to hold further consultations if the volume of imports of cereal substitutes were to increase dramatically. The EU secured a permanent 10% Community preference margin on the tariff equivalent calculations involved in the tariffication process. The final point was that the agreement included a guaranteed 5% minimum import access in all agricultural markets.

7.5.3 The Political Fallout in the EU, November 1992–December 1993

The Blair House Accord was a bilateral agreement and was not binding on the rest of the parties in the Uruguay Round. However, subject to some concessions to the EU in May and December 1993, it essentially defused agriculture within the round. The issue shifted from the international arena, the USA versus the EU, to internal to the EU: whether the EU member states would accept and honour this agreement made by their agricultural trade negotiators.

Soisson (the French farm minister) said in an interview with the FT (21 November 1992) that 'if we don't take a firm attitude on the GATT, the government could well fall'. The line taken was that the Blair House Accord was not *compatible* with the reforms of the CAP agreed in May 1992. An analysis of the substance and logic of this compatibility issue is given in Section 7.5.8. This period was characterized, as the Soisson quotation hints at, by internal opposition in the EU (focused on the Council of Ministers) led by French governments involved in the National Assembly elections of April 1993. The Blair House Accord, and how to oppose it, became an election issue. As Soisson said in the interview cited above, 'Agriculture still determines parliamentary majorities in France.' His estimate was that it determined 150 seats.

The Commission reached a 'grudging consensus' that the Blair House Accord was 'compatible' with the MacSharry reforms. This consensus was built on a ten-page document prepared by the Commission (SEC (92) 2267 (25 November 1992)).

Two successive French governments were unsuccessful in unravelling the accord (if that was ever their intention). However, concessions were won. Objections to the oilseeds part of the December 1992 agreement were assuaged by the CoAMs increasing the level of compensation for the set-aside requirements and production limitations of Blair House on 27 May 1993. From 1994, unit set-aside payments were to be 27% higher than envisaged in the reforms of May 1992. France won concessions on 7 December 1993 on the wider agricultural trade issues of Blair House. Some of these were significant. For example, the starting-volume figure, though not the end-volume figure, of subsidized exports was revised upwards. This gave the EU greater scope for exporting over the 6-year implementation period, making it easier to dispose of existing intervention stocks.

7.5.4 Interpretation of the Effect of the MacSharry Reforms on the Uruguay Round, 1990–1993

Section 7.5.5 looks at a brief model of how international negotiations are conducted and ultimately concluded. This is used in Section 7.5.6 to compare the De Zeeuw paper of July 1990 with the Draft Final Act of 1991. This Act, as described in Chapter 4, was the key agenda in the conclusion of the agriculture part of the Uruguay Round. Section 7.5.7 analyses the sequencing debate and Section 7.5.8 the issue of compatibility. Both are involved in the effect of the MacSharry reform process on the progress of the agriculture part of the Uruguay Round. Section 7.5.9 describes an assessment of the final Uruguay Round Agriculture Agreement. Section 7.5.10 is a conclusion to the discussion of the effect of the MacSharry reform process on the agriculture section of the round.

7.5.5 The Process of Compromise-defining

Disputes in the agriculture part of the Uruguay Round can be characterized as having been either qualitative or quantitative. A qualitative dispute is one in which the disputants fail to agree on a common ground for resolving the dispute because of the potential damage to certain political interests. The negotiating parties do not agree on the areas which need agreeing on. There is no substance to the negotiations; what is being offered is qualitatively different from what is being demanded. A quantitative dispute is one where there is common ground to the extent that each party is satisfied with the areas which require agreement. What is being offered by one side is connected with what is being demanded by the other; they are in the same units.

A continuum of possible compromises between disputants is possible only with a quantitative dispute. Qualitative dispute resolution requires discrete jumps to create a quantitative dispute before agreement can be finally reached. In a quantitative dispute the negotiations are centripetal; the disputants concede ground to the centre and towards the other parties. However, qualitative disputes are often centrifugal, because negotiating parties cannot agree on a common unit for negotiation. Each party has a tendency to put forward more extreme proposals to satisfy domestic constituencies in the knowledge that the proposals will never become the basis for serious international negotiations.

If it is imagined that the EU and the USA are in a quantitative dispute, their positions can be put at opposite ends of a continuum of feasible compromises. A compromise paper by GATT officials may be thought of as a third position lying on that continuum.

Of the three positions, the one prepared by the GATT secretariat is the most flexible: the secretariat is are not beholden to domestic vested interest pressures. To be successful as a broker in a situation of two separated and entrenched positions, the GATT secretariat position must be defined last because of this flexibility; that is, the ability to choose a point of compromise. This point on the continuum is chosen according to an assessment of the ability or willingness of each party to move to this position. It is here asserted that the very fact of promulgating a compromise point changes the bargaining party's point of maximum concession or willingness to move. This is clear because the last percentage conceded is not just a step nearer agreement as the previous percentage concessions were, but actually is the achievement of an agreement. More benefit is thus attached to the last steps, therefore the cost of those steps to the disputants can also be greater (if we assume some kind of rational agent making a decision to concede or not at the margin). Higher-cost choices are those further from the original starting-point of each negotiating party. The skill of compromise is judging that extra percentage each side will concede when agreement is near.

7.5.6 De Zeeuw and the Draft Final Act Compared

It is instructive to use this framework to compare the compromise paper of De Zeeuw and the Draft Final Act. De Zeeuw was commended by G7 leaders to their negotiators in agriculture at the Houston Summit of July 1990. It was to serve 'as a means of intensifying negotiations'. It lacked any detail about how much should be cut and how quickly; rather it tried to establish *what* should be cut. In this sense it attempted to move the agriculture part of the Uruguay Round from involving a qualitative dispute to involving a quantitative one. The kind of debate at Brussels in 1990 suggests that De Zeeuw was a failed attempt in these terms. The failure to establish a quantitative footing can be put down to a number of factors related to the intransigence of the EU and USA; see Chapter 6 for full details.

On the other hand, the Draft Final Act can be considered a successful piece of compromise setting because it informed Blair House, which in turn was the basis of the December 1993 final agreement for agriculture in the Uruguay Round. The Draft Final Act succeeded in defining the agriculture part of the round as a quantitative dispute. The effect of the MacSharry reforms on the ability to define a quantitative dispute in the Uruguay Round in 1992, as compared to the qualitative dispute which carried on through 1991, is considered in Section 4.2.6.

7.5.7 The Sequencing Debate

The sequencing debate arose from the parallel timetables of CAP reform and the Uruguay Round negotiations for the 18 months after the promulgation of the initial MacSharry plan in January 1991. Both had undefined endpoints, hence the question arose of which should be completed first.

The issue was resolved by the EU enacting the reforms of June 1992 mentioned above. As established in Chapter 6, the Commission's tactics were clear: the Commission wanted CAP reform first. The view taken by the MacSharry reform team was that if it was seen by the European agricultural constituency that CAP reform was a response to international negotiations then the domestic reform process would have withered. The politics of the CAP do not allow reforms which are seen as responses to USA pressure. MacSharry and his team freely admit that the reforms enacted in May 1992 had an effect on the progress of the agriculture part of the Uruguay Round. The official line maintained through the reform process by the Commission seemed to stretch credulity. MacSharry was quoted in the FT (22 May 1992, the day after the CoAM agreed the reforms) as saying, 'We can say, with our chests out and our heads held high, and with pride, that you [the USA] match what we have done, and then we can progress.'

A view opposite to that of the Commission was expressed by Peter Lilley in an interview with the FT (3 August 1991) in which he argued that a successful conclusion to the Uruguay Round must precede CAP reform. If the round waited for CAP reform it would suffer from the EU's 'limited capacity for flexibility'. He should have added that this was precisely the problem around November and December 1990. It is a matter of speculation and conjecture whether, had the Commission adopted this view of the proper sequence of domestic CAP reform and the Uruguay Round, (i) there would have been agreement on either, and (ii) any agreement would have been quicker or substantially different from what actually occurred.

7.5.8 The Compatibility Debate

This argument arose after the Blair House Accord of December 1992 and was the direct corollary of the sequencing arguments presented above. The debate turned on whether the commitments made at Blair House were 'compatible' with the CAP reforms of June 1992.

Prior to Blair House the French newspaper *Libération* printed a document leaked from an 'alleged' Commission source which shows that MacSharry had gone beyond CAP reform in GATT negotiations. Sources close to MacSharry quoted in the FT (16 November 1992) declared this report a 'fabrication'.

The Commission did produce a communication on the subject of compatibility (SEC (92) 2267 (25 November 1992)). It came to the conclusion – as the FT (27 November 1992) states, the only valid conclusion – that 'the most likely outcome' is that the exportable surplus will remain within Blair House limits. However, that conclusion was dependent on assumptions about consumption patterns, world prices and farm policy; for example, Rayner *et al.* (1993b) produce a conclusion that the EU may have to adjust policy to meet the reduction in the volume of subsidized exports in the cereals sector commitment.

The compatibility demand was used by France in its opposition to Blair House through 1993. This was a convenient political blocking tactic rather than a genuine issue. The MacSharry reforms of 1992 are a 3-year programme to 1995. Blair House is about agricultural policy over 1993–1999. The compatibility argument is asking whether the CAP, as it operated in 1995, was compatible with how it has to operate in 1999. Thus the EU has 4 years to make sure it is compatible. The question is a misnomer; only if farm policy is frozen for the 4 years after 1995 is there any issue at all. MacSharry exceeded his negotiating brief by talking about a timetable different from the reforms of May 1992, if his negotiating brief is taken to be those reforms and no further.

7.5.9 Assessment of the Uruguay Round Agriculture Agreement: its Effect on the CAP

This section provides a brief description of the changes made to the Draft Final Act after the MacSharry reforms of 1992 in order for agreement to be reached in the agriculture part of the Uruguay Round at Blair House in November 1992. These adjustments were in the level and content of domestic support reduction, the agreed reduction in the volume of subsidized exports and the negotiation of a peace clause. A consideration of the last-minute compromises involved in the Blair House Accord gives a lead into an assessment of the overall effect on the CAP of the Uruguay Round Agriculture Agreement.

Blair House was different from the Draft Final Act in explicitly exempting the EU's direct income compensation payments, introduced in the MacSharry reforms, from any reduction commitment. In GATT terms these payments, along with the USA's deficiency payments, were green boxed. This made compliance with the commitments in domestic support reduction agreed at Blair House 'virtually painless' (Ingersent *et al.,* 1995) for the EU and USA. Even before the MacSharry reforms, the EU had cut support price levels for some commodities relative to the 1986–1988 base period. The MacSharry reforms carried the price-cutting further, and, as noted above, the compensation payments for those price cuts were exempt from the AMS calculation.

At Blair House the EU and USA agreed a 21% reduction in the volume of subsidized exports compared to the 24% in the Draft Final Act. This reduction came as a result of EU pressure based on uncertainty over the effects of the MacSharry reforms on the volume of production and level of potential exports. However, Rayner *et al.* (1993b) suggest that for the cereals sector, the 21% reduction may still prove a constraint on the EU later in the implementation period and may demand some policy adjustments. The peace clause, as described in Chapter 5, was also an addition to the Draft Final Act after the MacSharry reforms of May 1992. This was part of the EU's strategy of setting the oilseeds regime of the CAP on a similar GATT footing to the cereals regime. Both regimes had been reformed in a similar way in May 1992.

7.5.10 Conclusion: the Effect of the MacSharry Reforms on the Uruguay Round

Ingersent *et al.* (1995) warn against the interpretation of the MacSharry reforms as a prerequisite for the Uruguay Round Agriculture Agreement. The Draft Final Act was published (in November 1991) before the final enactment of the MacSharry reforms (May 1992), although the authors hold that the Act was 'strongly influenced' by the prospect of the type of CAP reform proposed in COM (91) 258. The adjustments to the Act after the MacSharry reforms were largely the result of EU pressure in order to make Blair House compatible with the reformed CAP. These factors lead to the following interpretation of the relationship between CAP reform and GATT progress: 'rather than arguing that CAP reform paved the way to a GATT agreement, it seems more accurate to visualise the two sets of domestic and international negotiations pursuing parallel courses in the same direction before finally converging' (Ingersent *et al.,* 1995, p. 718)

If we use the simple model outlined in Section 7.5.5 the MacSharry reforms can be interpreted as moving the agriculture part of the Uruguay Round from involving a qualitative dispute to involving a quantitative one. The CAP reform proposals (COM (91) 258) moved the CAP in a way that reduced the number of qualitative disputes with the USA. The international agricultural trade negotiations were secured on a quantitative footing.

This chapter has established a link between the MacSharry reforms and the Draft Final Act. However, this book makes only limited claims about the nature of that relationship. Specifically, it is not claimed that (i) the effect of the MacSharry reforms on the acceptability of the Draft Final Act to the EU and the subsequent end of the Uruguay Round explains why the MacSharry reforms were enacted, or (ii) the effect of the MacSharry reforms in the Uruguay Round was an unintended consequence of reforms enacted for purely domestic reasons. Instead, it is noted that the MacSharry reform team admitted that elements of a CAP reform would inevitably influence the international arena of negotiations, and this chapter has aimed to describe that effect.

As noted earlier, the final Blair House Accord has had a limited effect on the post-MacSharry CAP. This fits into one of the central elements of this work: that CAP reforms enacted under the present institutional arrangements will always tend to be limited. Therefore, any international agreements that the EU enters into with regard to the CAP will similarly tend to be limited.

It is argued strongly in Chapter 6 that the causal link that involved the CoAM in deciding to enact CAP reform was not the demands of the Uruguay Round. However, MacSharry's reform team believed that some kind of domestic CAP reform would be necessary before the round could be completed. The need for a stronger position in the round was a motivation of the Commission for CAP reform. The important effect of a CAP reform in the international arena is not so much its details, but the fact that a reform has been made; the evidence presented above shows that there were last-minute adjustments for compatibility of the Draft Final Act with MacSharry, and such adjustments would have been possible for other types of CAP reform as well.

7.6 THE OILSEEDS DISPUTE

This section completes the history of the oilseeds dispute which was started in Chapter 4. The section describes the effect of the MacSharry reforms on the resolution of the dispute. Section 4.3.4 presented the USA strategy in international negotiations with the EU on the Uruguay Round and the oilseeds dispute in the summer of 1992 after the MacSharry reforms. This strategy was to accept the MacSharry reforms as a basis for a Uruguay Round agreement, but pursue demands for a limit on EU oilseeds production. The outstanding issue between the USA and the EU in bilateral negotiations on the agriculture part of the Uruguay Round in October and November 1992 was the volume of subsidized cereals exports. George Bush had pushed the deal in October against the backdrop of the US presidential campaign. The US pressure for a deal brought the two sides close but ultimately they failed to reach an agreement before the election. On 3 November (election day in the USA) in Chicago, USA Agriculture Secretary Edward Madigan and MacSharry were close to agreeing a deal which included the commitment to reduce the volume of subsidized farm exports by 21%. After the volume of subsidized exports, there were two outstanding issues: the USA demand for a 10 Mt ceiling on EU oilseeds production against the EU's proposal of some formula based on the area of crops planted, and the issue of a peace clause. The EU was demanding that both sides renounce the future right to challenge the other's cereals regimes before the GATT panel.

During the negotiations (a comprehensive account of the details is given by Grant, 1995b) Delors telephoned MacSharry and informed him that an agreement of a 21% reduction in the volume of subsidized farm exports would go against his negotiating mandate and the CAP reforms agreed in May 1992, and stated that he would oppose it in the Commission. MacSharry thought that 21% had been agreed before he left for Chicago. The US team sensed the pressure that MacSharry was under and thought that the EU would not be able to agree on 21%, so they withdrew their offer and rejected EU demands for a peace clause.

MacSharry reflects in an interview with Grant (1995b, p. 10) that without Delors's pressure, 21% would have been agreed and no pressure would have been applied by the USA on the peace clause issue. In such circumstances MacSharry believes a deal would have been completed and he could have 'told Delors to stuff it' (Grant, 1995b, p. 175). In fact, MacSharry announced his resignation from GATT negotiating duties, complaining of pressure from Delors. He went to the press with his assertion that Delors was looking after French interests (FT, 5 and 6 November 1992). Delors was reported to be resentful of what he thought was a campaign against him by Andriessen and MacSharry (separately) in the press.

The USA responded immediately to the breakdown of bilateral negotiations by threatening tariffs of $300m on EU products, effective from

5 December, if some kind of acceptable agreement was not reached on the oilseeds dispute (the only part of the negotiations on which it could legally make such a threat). This threat hung over the EU as it attempted to sort out a position with regard to resolving both the oilseeds dispute and the wider agriculture issue in the Uruguay Round.

Although this book concentrates to an extent on the Delors–MacSharry relationship, MacSharry and Andriessen (the latter had overall responsibility for GATT negotiations throughout MacSharry's time) disagreed strongly. There was tension between the Andriessen position that issues in agriculture could be conceded in the interests of a wider agreement, and MacSharry's tough line that agriculture (his portfolio) was most important and each concession had to be resisted. Andriessen abstained on COM (91) 100 in the College because he believed it did not leave enough room to complete the Uruguay Round.

Initially, Delors stood apart from this tension at the centre of the EU's negotiating stance in the Uruguay Round. Grant (1995, p. 173) says that 'Delors did not discourage a deal: he was happy to let Andriessen tackle GATT and make a mess of it. Delors never seemed to make GATT a priority.' However, the chaos and damage to the EU's international credibility after Chicago prompted Delors to intervene. This application of political influence brought MacSharry back to the GATT negotiating team, who along with Andriessen and Delors himself completed the negotiations for the agriculture section of the Uruguay Round (at Blair House on 20 November 1992).

7.6.1 Final Agreement at Blair House

The oilseeds and the wider agriculture negotiations of the Uruguay Round were settled at Blair House. The three central points were: (i) a 21% reduction in the volume of subsidized exports over 6 years; (ii) a peace clause, agreed to be valid for 6 years; (iii) an agreed 5.13 Mha area limit on oilseeds production in the EU.

The 5.13 Mha limit on oilseed plantings applied to the three major oilseeds (rape, sunflower and soya) from the 1995 harvest. The EU agreed to reduce its plantings by a minimum of 10% from this base area of 5.13 Mha over the 6 years. Further, there was agreement that EU set-aside in the oilseed sector would be the same level as the overall arable set-aside but never less than 10%. Set-aside for other arable crops of the MacSharry reforms was 'aggregative'; that is, of all arable land a certain percentage of land must be set-aside. All different crops which can be grown on arable land must be added together for the purposes of set-aside. The Blair House Accord stated that a specific percentage must be applied to the area for growing oilseeds. The EU was also given a limit to the growth of oilseeds for non-food uses.

The oilseeds dispute was resolved by agreement of a quantitative restriction on the growth of the EU oilseeds industry. This bears little relation to the legal basis of the dispute and ignored the two existing GATT rulings against the EU.

7.6.2 Reaction to the Blair House Accord

The Blair House agreement by the agriculture GATT negotiating team of the Commission found virulent opposition in the Council of Ministers from France. Various veto threats were made and invocations of the Luxembourg compromise. The Bérégovoy government maintained opposition to the deal until it was defeated in the National Assembly elections of April 1993. The new government was becalmed on the oilseeds part of Blair House by increased set-aside compensation (benefiting all EU farmers) and the allowance of growing of certain crops (rapeseed, sunflower) on set-aside land for non-food uses (e.g. methyl ester, a diesel substitute) agreed by the CoAM on 27 May 1993. There was previously very little margin in this business at farm level, but, as *AgraEurope* argued, the oilseeds dispute was settled by the EU raising the profitability of growing oilseeds for biofuel 'from the very marginal to the comfortable'. French opposition to the wider Blair House Accord rumbled on to December 1993. This is where *AgraEurope* presumably sees the end of the oilseeds dispute.

7.7 CONCLUSIONS

This chapter has considered the economic and financial effects of the MacSharry reforms. In addition, it has also continued the theme that CAP reforms are the outcome of a process by considering the effects of the MacSharry reforms on the circumstances and pressures which gave rise to their enactment.

The *ex ante* analyses presented predict EU agricultural production falling in the medium term in response to the reduction in support prices, higher farm incomes and higher EAGGF expenditure as DIPs outweigh the reduction in the budget cost of export subsidies. Actual EU cereals production levels in 1995 were the same as in 1993 despite an increase in yields. This was due to a decline in the area cultivated in response to the set-aside requirements. The EAGGF budget has remained within its guideline in 1994, 1995 and 1996. However, this has more to do with favourable trends in world prices than the direct effect of the MacSharry reforms (see Section 7.3).

The MacSharry reforms influenced the final Uruguay Round Agriculture Agreement. This claim falls short of saying either: (i) that this effect was the reason why the MacSharry reforms were enacted, or (ii) that this effect was an unintended consequence for the Commission of the enactment of the MacSharry reforms. The oilseeds dispute was drawn into the Uruguay Round negotiations and settled as part of the Blair House Accord. The circumstance of politically difficult automatic price cuts has been removed by the dismantling of the stabilizer regime.

The final step in this chapter has been the discussion of the stability over time of each member state's net pay-off (budget plus preferential trade effects) from the CAP. Like the previous reforms of the CAP in 1984 and 1988, the MacSharry reforms have not significantly affected the net position of most member states. This point about stability in the CAP is picked up in Chapter 9.

Chapter 8

Interpretations of the MacSharry Reforms

8.1 INTRODUCTION

Chapter 5 selected the institutions framework as a basis from which to construct evidence concerning the MacSharry reform process. This selection was made on the basis that the institutions framework provided a better interpretation of the reforms of the CAP in 1984 and 1988 than either the interest groups or prominent players frameworks. This chapter seeks to assess whether the MacSharry reform process is better interpreted using the institutions framework. Each of the frameworks is considered in turn and is judged according to how the evidence of the MacSharry reforms constructed in Chapters 5, 6 and 7 can be understood using that framework.

An interpretation of the MacSharry reform process will provide the substance of the answers to the three main questions of this book: why did the MacSharry reforms occur, when did they occur and why did they occur in the way that they did? Separate sections of the chapter consider each of the rival frameworks, and one is chosen, in terms of providing the clearest, most coherent and most concise interpretation of the evidence of the MacSharry reform process. This interpretation is used in Section 8.5 to compare MacSharry with the reforms of the CAP in 1984 and 1988 (the histories of which are detailed in Chapter 3). The understanding of the MacSharry reforms will be improved by a consideration of whether the elements of the MacSharry reform process highlighted by the chosen framework as important in that process are different or similar in previous reforms of the CAP.

8.2 THE INTEREST GROUPS FRAMEWORK

A criticism of the interest groups framework presented in Chapter 5 was that its application to the question of explaining changes in agricultural policy tended to be too abstract and, in particular, not specific to the EU. The framework rests on the claim that agricultural policy is a function of the balance of interest group power, and therefore changes in agricultural policy are the result of changes in that balance of power. There are two levels of government in the CAP decision-making system: the EU institutions and the national governments. Therefore, the question becomes whether there is evidence at either of these different levels of government reflecting a shift in the balance of interest group power during the MacSharry reform process.

As Becker (1983) notes, the power that an interest group enjoys can be affected in two ways: by a change in the political resources at its disposal, and by a change in the pay-off to each unit of political resources employed in seeking some government action. The evidence constructed in Chapter 6 of the causal links at work in the MacSharry reforms does not seem to show a changing balance of interest group power (by either means) in favour of consumers and against taxpayers (the shift in the burden of agricultural policy brought about in the MacSharry reforms). None of the fieldwork interviews conducted as part of the research for this work produced any evidence of a shift in the overall balance of interest group power when compared with the previous reforms of the CAP in 1984 and 1988. This includes interviews at COPA, at the NFU and with government officials at EU and national level. The main conclusion of the research conducted for this book in this area is that interest groups were not a significant factor in the type of reforms proposed by the Commission or the timing of those reform proposals.

The Commission, in proposing the MacSharry reforms, was responding to pressures for a reform of the CAP. These pressures to initiate CAP reform did not come from interest groups, nor were interest groups the 'transmission' mechanism from some objective situation affecting European agriculture to demands for the Commission to introduce proposals to change the CAP.

However, the scale of agricultural policy interest group lobbying is such that the conclusion of this book that interest groups were marginal in the initiation of the MacSharry reform process should be tempered. Interest groups do have a role in CAP decision-making and could be more significant in informing future reforms of the CAP. As Hull (1993) and Egdell and Thomson (1997) note, there has been a proliferation in the number of interest groups in Brussels and the extent of lobbying activity surrounding the development of the CAP. Pedler (1994) estimates that there are approximately 10,000 lobbyists in Brussels. The European Commission lists 637 pan-European non-profit-making organizations with which it deals, 118 of which (19%) are agriculture or food-related (Commission of the European Communities, 1996). Just over half of these are concerned with food products and processing, and a fifth with trade in agricultural and food products. Egdell and Thomson (1997, p. 2) say, 'one can estimate that around £100 million must be spent annually on salaries alone of agricultural lobbyists'. The larger farming and commodity groups have frequent contact with the Commission owing to their presence on management and advisory committees. Interest groups with this level of contact and lobbying activity are important providers of information. They may enjoy influence as a potential alternative source (to the Commission) of quantified data about the CAP and its effects.

8.3 THE PROMINENT PLAYERS FRAMEWORK

The prominent players framework holds that CAP reforms and the process which accompanies their enactment should be analysed in terms of the interaction of prominent players. These prominent players are the institutions of the state and interest groups at both national and EU levels. Their relationships determine the policy process. Chapter 5 argued that the relevant category of state–group relationships for the CAP decision-making system was a policy community. Whether the prominent players framework is applicable to the MacSharry reforms depends on: (i) the existence or otherwise of a policy community at EU level, and/or (ii) the existence or otherwise of a series of policy communities at national level.

There was only limited evidence in Chapter 6 that the MacSharry reform process displayed the characteristics of the operation of a policy community at an EU level. COPA was not consulted on the decision of the Commission to propose a CAP reform or the substance of the reform proposals. In the first instance, COPA had no agreed response to the type of reforms proposed or any proposals of its own. Subsequently through the reform process, COPA failed to agree a common line with regard to the MacSharry reform proposals. In policy communities, as described in Chapter 5, members interact across all aspects of policy and there is a high degree of consensus as to the means and ends of policy. This was not obviously the case in the MacSharry reform process.

The second way in which the prominent players framework may be relevant is if national agricultural policy interest groups are significant in affecting the positions of member states with regard to CAP reform; that is, there exists some kind of national policy community. Chapter 5 sets up a number of criteria by which to judge the significance of this influence: the cohesion of the farm lobby, the functional relationship with the national minister of agriculture, the importance attributed to agriculture by the national government, the political power of the minister of agriculture, whether the farm lobby can control the agenda of policy or enjoys a level of political resources which are necessary for the enactment of a CAP reform.

The relationship between agricultural policy interest groups and the positions of member state governments in the MacSharry reform process in the UK, France and Germany is considered here. These are the three most (politically) important governments involved in CAP reforms.

The NFU in the UK is an example of a cohesive farm lobby – it represents about 80% of UK farmers. However, its current relationship with MAFF is at an 'arm's length' (interviews); it competes for the agenda of MAFF with an increasing number of other groups (Cox *et al.*, 1986; Winters, 1987). Further, MAFF itself is not a politically powerful department within the UK government and very often the agriculture minister has to adopt the Treasury line in CAP negotiations; that is, focus almost exclusively on the budget consequences (at EU or UK level) of CAP decisions. Further, the evidence of the Treasury's domination of the UK government's position with regard to the CAP can be seen in the willingness to be outvoted in the CoAM instead of accepting a particular outcome as inevitable and within that constraint seeking the best compromise deal for the UK agricultural interest. The NFU has only a limited effect on the position of the UK government in CAP negotiations and it is difficult to claim that a policy community exists on this issue.

The influence of the French farm lobby on the position of the French government during the reform process was limited by its lack of cohesion; there is no single dominant farm organization as in the UK case. Although the Minister of Agriculture is a much more politically important post in the French government than in the UK one, there is no evidence that the reaction of French agricultural policy interest groups significantly changed the French government's perception of the national agricultural interest. As described in Chapter 6, the French government's position reflected a split in the French agricultural sector between an internationally competitive part and smaller uneconomic farms which rely for their survival on some kind of state support. Different agricultural policy interest groups represent different sections of French agriculture. These groups had different reactions to the MacSharry reforms. However, none of them was able significantly to influence the agenda of the Commission or the options facing the Minister of Agriculture, Mermaz.

The French government's initial reaction to the MacSharry reform proposals was consistent with the ambiguity towards the CAP displayed by successive French ministers of agriculture. The reform proposals of the Commission were neither formally rejected or accepted. Mermaz consistently expressed concern that the size of the cuts in support process would significantly reduce the level of export subsidies that the largest and most efficient French farms received. At the same time, he responded positively to a lot of the rhetoric MacSharry himself used in the CoAM to describe the reform proposals; in particular, the emphasis on preserving the maximum number of farmers in Europe. This conflict in the French position was resolved only at the end of the reform process after the German government had shifted to the position of positively arguing for the need to agree a CAP reform before the 1992/93 price package. This was the only time during the reform process that the French government was unambiguously positive about the need to agree CAP reform along the lines proposed in COM (91) 258. This combined Franco-German position (see below and Chapter 6) was a necessary factor for agreement in the CoAM on the MacSharry reforms.

The movement in French position to accepting the reforms is one thing that requires to be explained to understand the MacSharry reforms. If French agricultural policy interest groups are to be prominent players in the MacSharry reform process then they must to some extent have affected the shift in French position from being non-committal to supporting in the CoAM the reform proposals contained in COM (91) 258. The evidence presented in Chapter 6 does not support the idea of French agricultural policy interest groups as prominent players; the shift in position by Mermaz was heavily influenced by the movement in the German position described below and the operation of the stabilizer regime producing politically unpalatable price cuts for 1992/93. These factors exist outside the dynamics of any French agricultural policy community.

A similar pattern emerges in the relationship between German agricultural policy interest groups and the German federal government. All the interview results showed that the relationship between the German farm unions and the German government exhibited the first three factors listed by Petit *et al.* (1987): a cohesive farm lobby with a strong functional relationship with the Ministry of Agriculture and a politically powerful minister of agriculture. If a policy community can be said to exist anywhere in the CAP decision-making system it is here. However, it should be noted that there has been some work suggesting that the lobbying power of the Deutsche Bauernverband (DBV) has waned slightly since the mid-1980s (e.g. Von Cramon-Taubadel, 1993).

The decision by Keichle to move position on the MacSharry reforms (in March and May 1992, described in Chapter 6) was conditioned by the options he faced and the environment which forced a choice to be made. The agenda facing him in May 1992 had two options: uncompensated price cuts under the 1988 stabilizer regime, or compensated price cuts with the MacSharry proposals. Both options broke Germany's historical position with regard to the CAP, which can be summarized as resistance to any proposals to reduce nominal support prices. The choice of the MacSharry reforms reflected the course with the lower political costs. German interest groups could not in any sense influence the choice presented to the German agriculture minister at the May 1992 CoAM; they were external to any national policy community.

Further, the attitude to the choice presented was consistent with the historical position of German governments in CAP negotiations; there was no shift in position by the German government in this sense. There was no shift in the national policy community which affected the German position; the movement by the government to supporting the MacSharry reforms (which is central to understanding the reform process) was conditioned by the options and the need to make a choice.

The influence of agricultural policy groups and whether they are part of a policy community is highly variable across time and across countries. Even where they are included in a policy community, these are nationally based and membership does not equate to being a prominent player in the CAP reform process. Agricultural policy groups are not directly involved in the CAP reform process and do not significantly help in the understanding of that process. The prominent players are the Commission and the member state governments, in the CoAM or the European Council. This discussion seems to suggest that the institutions involved in the CAP decision-making system are much more central to understanding the operation of that system; these are the prominent players.

The step to the institutions framework is as follows. The prominent players of the CAP policy network, the institutions, do not form a policy community. The institutions of the CAP decision-making system exist in a state of tension with each other; over aims, means, and levels of agricultural support. Their interaction is neither regular nor consistent. Information about the CAP is controlled by DG VI, which is organized internally by horizontal separation of functions by commodity. This leads to a tendency for cliques and cabals to form to an extent where it may be doubted whether there is a community even within an individual institution.

As with the interest groups framework, the claim that the prominent players framework seems to provide only a limited insight into the MacSharry reforms must be considered alongside the possibility that at other times the relationships which establish certain agricultural policy interest groups as prominent may be important. There is inertia in the CAP decision-making system. Therefore, an agricultural policy interest group which became prominent in exceptional circumstances might retain influence and power in the CAP decision-making system for some time afterwards.

8.4 THE INSTITUTIONS FRAMEWORK

This section presents the institutions framework as the most appropriate of the three rival frameworks outlined in Chapter 5 for organizing and interpreting the evidence of the MacSharry reform process. Each of the institutions involved in the decision to reform the CAP has a different agenda and is involved at different times. This temporal separation and separation of agendas means that the institutions of CAP decision-making can be interpreted as existing in a state of competition with each other on the issue of

CAP reform. This also fits in with the notion of CAP reforms being the result of a process.

The institutions framework starts the analysis from the point that CAP reform proposals originate in the Commission and are formally enacted in the CoAM. These are the two fixed points in a CAP reform process. The MacSharry reforms are an example of the constraints, motivations and requirements existing in the CoAM being clearly different from those of the Commission (see Chapters 5, 6 and 7). This can be interpreted as an example of inter-institutional competition.

In January 1991, the Commission required a new position in the agriculture part of the Uruguay Round of the GATT negotiations and also felt that the budgets of certain market support regimes had reached a critical level; that is, they were threatening the 1988 budget guidelines. The reason the Council of Agriculture Ministers enacted the reforms at its May 1992 meeting was because of the choice presented: an uncompensated price cut of 11% (*Agence Europe,* 11 April 1992) under the stabilizer regime, or the MacSharry reforms with much larger, but compensated, cuts in CAP support prices.

The tension between the objectives of the Commission and what the CoAM will agree to is a major theme in the institutions framework. In 1984, 1988 and 1992 the CoAM was an impediment to the Commission's agenda for CAP reform. In none of these cases did the CoAM actually change substantially what the Commission originally proposed. The CoAM enjoyed only the power to obstruct, to the extent that the reform process was reduced to glacial speed. One of the results of my fieldwork interviews is the admission by MacSharry and members of his reform team that had they known in advance how difficult the reform process during the period 1990–1992 was going to be they would not have started it (see Chapter 6). The existence of this threat of intransigence allows the CoAM to try to restrict the ability of the Commission to make policy-shaping decisions and affects the reform process.

The idea of the CAP reform process as a series of competing institutions does not apply merely between institutions but also within them. The Commission is affected by competition between different Directorates-General for control of the CAP. DG VI (Agriculture) has in recent years faced competition from in particular DG II (Economics) and DG XIX (Budget) for control of the agenda of European agricultural policy. Also, the separation of DG VI by commodity division stifles the progress of reform ideas within this Directorate-General. Hence Commissioner MacSharry used the tactic of working in small, informal and *ad hoc* teams, as described in Chapter 6, to develop his reform plans. This circumvented the main policy development routes of DG VI. Further, with the support of Delors, the College of the Commission was presented with the full reform proposals only very late on in the construction of the plans. MacSharry's ability to navigate his reform plans through the different competitions within the

Commission for control of the CAP reform agenda can be attributed, in part, to political leadership.

The CoAM is more obviously subject to internal competition; it contains a series of competing national agendas for the benefits of CAP expenditure. The institutions framework encourages focus on the options faced by individual members of the CoAM. There is some point at which the political costs of agreeing CAP reform are less than those for maintaining the status quo. As described in Chapter 6, in the MacSharry reforms this point came when the political costs to the French and German governments of the continued operation of the stabilizer regime became greater than the political costs of agreeing the MacSharry reforms.

The political costs in the MacSharry reform process arose from cuts in the nominal support prices. In the CoAM, the key issue was the compensation of those price cuts. The movement in position, which removed the impasse that had existed in the CoAM since July 1991 and the COM (91) 258 proposal, came from Germany. Germany's consistent line since the inception of the CAP had been to defend the nominal level of support prices. As described earlier, the options at the May 1992 CoAM were compensated or uncompensated cuts in nominal support prices. Both options broke the historical line of German governments, but the compensated price cuts had lower political costs for Keichle. Hence the German government supported the MacSharry reforms and the reforms were enacted.

8.5 MACSHARRY AND PREVIOUS REFORMS COMPARED

Section 8.4 uses the institutions framework to give an interpretation of the MacSharry reform process. This section uses that interpretation to make some comments about the extent to which the MacSharry reform process was different from the processes which accompanied the reforms of the CAP in 1984 and 1988.

The political costs in the reforms of 1984 and 1988 arose from different sources as compared with those of the MacSharry reforms. Political costs motivate individual CoAM members to agree reform proposals. In the milk quotas and stabilizer reforms, these political costs were generated in the CoAM through competition between Ecofin and the CoAM for control of the CAP. The operation of the CAP has effects outside the constituencies of the members of the CoAM. This 'spillover' into the policy area (or turf) of another Council creates competition for control of the CAP. This takes the form of an attempt by another Council to influence the CAP decision-making system in order to limit the effect of CAP decisions on that Council's policy areas. In 1984 and 1988, Ecofin was politically strong enough in both cases to force the issue of the CAP budget to the European Council. It was the European Council which subsequently enacted the CAP reforms of 1984 and 1988.

The conflict for control of the CAP can in certain circumstances led to the European Council being involved in CAP decisions as an arbiter. The issue of CAP reform reaches the European Council in an atmosphere of crisis provoked by competition between various Councils; in other words, inter-institutional conflict. The European Council is much more disposed to reform the CAP than is the CoAM. Moyer and Josling (1990) note, with reference to the CAP reforms of 1984 and 1988, that except for the UK and Dutch representatives, each CoAM member was motivated almost exclusively by the farm interest when considering the reforms. The European Council weighs the agricultural interest against other EU interests. The bias of the CoAM in focusing exclusively on the benefits to the different national agriculture sectors was removed from the decision-making process by the intervention of the European Council.

The European Council was not involved in the MacSharry reforms. Further, there was not the atmosphere of crisis which accompanied the two reforms of the CAP in the 1980s. The political costs which forced the CoAM to agree the reforms in May 1992 were generated by the stabilizer regime implying cuts in nominal support prices. The incidence of these political costs was directly on the members of the CoAM rather than indirectly either through pressure from other Councils or being effected by the intervention of the European Council. However, the MacSharry reforms were consistent with previous reforms in that they were enacted because of short-term political costs impacting on the myopic sights of the members of the CoAM.

Chapter 3 explains how the institutional arrangement of the EU budget was changed in 1988 to counter the ability of the CoAM to act without regard to the financial implications of its decisions (this ability came from the fact that the budget agreement and the farm price agreement occur at different times of the year – there was no equivalent of the UK's 'Star Chamber'). The 5-year budget guidelines agreed in 1988 and 1992 were designed to control the rise in the budgetary cost of the CAP. The 1988 reforms were part of a wider budget package inspired by the plans of Delors. The MacSharry reforms can similarly be interpreted as part of a competition between parts of the Commission and the CoAM for control of CAP expenditure.

Fearne (1991) and Chapters 1 and 2 of this book describe how the CoAM, through the mechanism of the annual price review, had incrementally increased nominal institutional support prices on an annual basis through the history of the CAP, affecting both its cost and its overall direction. The partial shift from price support to income support through DIPs means that the importance of the annual price review has been downgraded; the agreement of the MacSharry reforms meant that support price levels and DIPs levels were agreed for 3 years, reducing the influence of the CoAM.

8.6 CONCLUSIONS

Chapter 1 proposed a distinction between turbulence and stability in the CAP. Specifically, reforms of the CAP should be understood as turbulence against a long-run stability in its operation and effects. With reference to the MacSharry reforms the framework chosen is required to: (i) provide an interpretation of the reform process, the 'turbulence', and (ii) support an explanation of the limited nature of CAP reforms, the 'stability'.

The institutions framework has been chosen as providing the most cogent interpretation of the evidence of the MacSharry reform process gathered for this book and presented in Chapters 4, 6 and 7. The institutions framework says that CAP reforms start in the Commission; the factors which affected the Commission through 1990 are the reasons why the MacSharry reform process started. These are described in Chapter 6. The progress of the reform proposals through various stages depends on the internal balance of the CoAM, the president of the CoAM and the willingness of the Agriculture Commissioner and the College to make concessions on their original proposals. The causal links involved in these relationships are given in Chapter 6. The reasons why the reform process ended when it did and in the way that it did can be found by analysing the internal dynamics of the CoAM, which until 1995 was composed of 12 national Ministers of Agriculture. Chapter 6 describes the crucial interactions in the final agreement of the MacSharry reforms as occurring here.

This chapter maintains that the relationship between the Commission and the CoAM holds the key to understanding turbulence in the CAP; research should concentrate on competition for control of both institutions, the relationship between the two institutions, and the relationship between these two and other institutions. These dynamics can be used as a framework for collecting evidence about the causal links at work in CAP reforms.

The public choice paradigm starts from the point that public decisions are taken by individuals and adds to this the premise that the perspective and behaviour of these individuals are conditioned by their place in the CAP decision-making system (see Chapter 5). Chapter 6 shows that the individuals involved in the decision to enact the MacSharry reforms were in the Commission. It is the contention of the institutions framework that the views of the individuals involved in the MacSharry reform process were conditioned by their being members of particular institutions in the reform process.

The three frameworks of Chapter 5 are suggested structures of the CAP decision-making system. The public choice paradigm assumes that individuals fit within these structures. The structure–individual agent relationship is discussed further in Chapter 9. It is an important point for how these different frameworks can be employed to understand a CAP reform process. Although this work has assumed a direction of causation from structure to individual behaviour, in certain instances this direction may be reversed. An individual may enjoy the political skills and influence to alter the structure of the CAP decision-making system. For example, a strong Agriculture Commissioner may be able to impose a 'prominent players'-type structure. An effective COPA leadership might mould the decision-making system towards an 'interest groups'-type framework. This book holds that the number of circumstances in which individuals could affect the structure is limited. However, the possibility exists, and this constrains the degree of bias towards the institutions framework declared in this book.

Chapter 9

Conclusions

9.1 INTRODUCTION

There are two complementary objectives in this book. The first is to understand the MacSharry reforms. Understanding has been defined (in Chapter 1) in terms of answers to three questions: why were the MacSharry reforms enacted, when were they enacted and why were they enacted in the way that they were? The complementary objective is a more general understanding of how the CAP decision-making system operates. The specific evidence of the MacSharry reforms contributes to the arguments about stability and turbulence in the CAP.

Sections 9.2 and 9.3 provide a summary of the claims of this book with regard to the first objective. Section 9.3 notes the differences between the MacSharry reforms and the milk quota reforms of 1984 and the introduction of the stabilizer regime in 1988. Section 9.4 considers how the points made in Section 9.3 affect the claim that the institutions framework is the most convincing of the three analytical frameworks for CAP reform set up in Chapter 5 for a general understanding of CAP reforms. Section 9.4.1 considers how the role of individuals may be understood within the common analytical frameworks of the CAP decision-making system. This is the complementary objective in this work. The final section of the chapter links the main conclusions of the book with the political economy literature in agricultural policy.

9.2 UNDERSTANDING THE MACSHARRY REFORMS

9.2.1 Why Were the MacSharry Reforms Enacted?

This answer to the question of why the reforms were enacted requires a description of the pressures and causes of the MacSharry reforms. As is emphasized throughout this work, CAP reforms are the result of a process. Hence any description of the causes of the MacSharry reforms must be divided into two: those causes which triggered the reform process and those causes which operated at its conclusion. To state this another way, there are causes which operate at the beginning and the end of element 2 of the CAP decision-making system (see Chapter 1), and these are not necessarily the same.

The main cause of the start of the MacSharry reform process was the combination of MacSharry as Agriculture Commissioner and Delors as Commission president after 1989. They shared an ambition for CAP reform. The background to this ambition is described in Section 6.1. Sections 6.9–6.11 describe the major cause of the enactment of the MacSharry reforms in May 1992 (the end of the reform process) as being the operation of the stabilizer regime at that time. In particular, the automatic price cuts implied by the regime at the time of the negotiations of the 1992/93 price package, if accepted, would have imposed substantial political costs on the members of the CoAM.

9.2.2 Why Were the MacSharry Reforms Enacted When They Were?

The previous subsection provides the context of the causes of the MacSharry reforms by summarizing the account given in this work of the timing of the MacSharry reform process. Section 6.1.4 details how the start of the process was heavily influenced by the deadline of December 1990 for the conclusion of the Uruguay Round of GATT negotiations. The collapse of that round at Brussels in early December 1990 amid USA accusations of EU intransigence, was followed a week later by the start of the campaign by MacSharry and his team to persuade the College to agree their CAP reform proposals. Section 6.1 outlines how the international dimension affected the timing of the start of the domestic policy reform process. The timing of the conclusion of the reform process was tied to the need to agree the 1992/93 price package and the implications for that agreement on the operation of the stabilizer regime. As described in Chapter 6, the stabilizer regime implied an automatic cut of 11% in nominal support prices. It was the effect of this on the political calculations of members of the CoAM when they came to agree the 1992/93 price package in March and April 1992 which led to the agreement of the MacSharry reforms.

9.2.3 Why Were the MacSharry Reforms Enacted in the Way That They Were?

The CAP reform process 1990 to 1992 was a response to a set of circumstances. A full answer to this question requires an account of the substance of the MacSharry reforms; given the causes (Section 9.2.1) and the set of circumstances (Section 9.2.2) of the reform process, why was this type of reform proposal made by the Commission and ultimately enacted by the CoAM?

COM (91) 100 (Commission of the European Communities, 1991a) was heavily influenced by Commissioner MacSharry and his perspective on the CAP – in particular, his views on the CAP's purpose and what its main failings had been. This personal perspective is described in Section 6.1. The main differences between COM (91) 100 (agreed by the College on 31 January 1991) and the MacSharry reforms (agreed by the CoAM on 22 May 1992) were first, a removal of most of the elements of modulation in the different compensatory payments introduced, and second, a reduction in the level of cuts in nominal support prices. The removal of most of the elements of modulation initially proposed was the result of the stance taken by the UK government in the CoAM and to a lesser degree the influence of the Danish delegation. The reduction in the level of price cuts was heavily influenced by the negotiating position of the German government.

9.3 DIFFERENCES BETWEEN MACSHARRY AND THE REFORMS OF THE CAP IN THE 1980S

At the most obvious level, the MacSharry reforms were different in substance from the reforms of the CAP in 1984 and 1988. The implementation of substantial cuts in institutional support prices, the introduction of DIPs as compensation and the linking of compensation payments to compulsory set-aside requirements all challenged existing notions of what was a 'politically feasible' CAP reform (see Hagedorn, 1985, for the argument that DIPs were politically impossible).

The MacSharry reforms were different in having an international dimension. This affected, at least, the timing of the start of the reform process (Section 6.1). The Uruguay Round including a section reducing agricultural support levels, was due to be concluded at Brussels in December 1990. Chapter 3 describes how the pressures of the international dimension were 'inconsequential' in both the reforms in the 1980s.

The causes of the MacSharry reforms are also different. As noted in Section 9.2.1, it was the *toughness* of the stabilizer regime which eventually forced the CoAM to reach an outcome on the reform proposals and conclude the MacSharry reform process. Toughness is defined here in terms of the political perspective of members of the CoAM. The CAP would be operating in a tough manner if it imposed political costs on members of the CoAM such that they would contemplate agreeing CAP reform. In 1984 and 1988 the then current CAP was not tough from the point of view of members of the CoAM; rather, its operation was creating effects which had an incidence on other institutions of the EU. The CAP was exhausting the budgetary resources of the EU. Other institutions responded to this spillover effect of the CAP by applying pressure on the CoAM to reform the CAP. This inter-institutional tension was the cause of the reform of the CAP.

This difference in causes finds a reflection in the institutional interaction in the MacSharry reform process compared with the reforms in 1984 and 1988. In the latter two instances the European Council was centrally involved in the decision to reform the CAP. It was involved in an *arbitrator* role when the spillover effect from the operation of the then CAP was the potential exhaustion of the budgetary resources of the EU. This fostered inter-institutional tension which required the intervention of the European Council. In the MacSharry reforms, the European Council was not involved and there was not the sense of crisis characteristic of the reforms in the 1980s. This reflects the fact that it was not the operation of the CAP creating a budget 'spillover' effect from the CoAM to other institutions which created the pressures for reform. Instead, it was the toughness of the CAP in 1992 which forced the CoAM directly into adopting the MacSharry reforms in May of that year.

The MacSharry reforms were different for MacSharry as a political personality. His relationship with Delors and their ability to drive a CAP reform agenda through both the Commission and the Council (detailed in Chapter 6) represented a difference as compared with the reforms of 1984 or 1988. Their CAP reform agenda had a distinctive personal element, which was not the case with Dolsager in 1984 or Andriessen in 1988. MacSharry's beliefs about what was required in a CAP reform were not a response to a deteriorating budget situation; rather, they were founded on his experience in agribusiness and the politics of agricultural policy.

However, the differences between the MacSharry reforms and reform process and the reforms in the 1980s having been noted, it is worth highlighting the major similarity. This was drawn out in Chapter 7 on the effects of the MacSharry reforms: if a radical CAP reform is one that disturbs the balance of pay-offs from the CAP at the member state level, then none of the reforms cited here is radical. The MacSharry reforms, though different in elements of substance and in the process of their enactment, were not radical; they did not upset the fixed political bargain which exists in the CoAM.

9.4 THE INSTITUTIONS FRAMEWORK

This section considers whether the differences between the MacSharry reforms and the previous reforms of the CAP can be accommodated within the institutions framework, a question examined in Chapter 8. The conclusion was that the evidence of the MacSharry reforms did not affect the claim made in Chapter 5 that the institutions framework provides the most cogent and concise interpretation of CAP reforms.

In particular, the institutions framework provides an insight which allows an account of the one factor that has been unchanged through each reform of the CAP. The outcome of CAP reforms, expressed in terms of the national distribution of CAP pay-offs, is relatively stable across time. Section 1.5 sets up the distinction between stability and turbulence in the history of the CAP. CAP reforms have represented turbulence in the context of long-run stability in the CAP. Using the definition of 'radical' from Section 1.5, none of the reforms of the CAP has been radical. The institutions framework provides a double insight into this observation.

First, the CoAM will only reach a decision on the basis of short-term political costs. This is because its membership consists of short-term political appointees whose interest in agricultural policy is generally short lived (though there are a number of notable counter-examples). Any CAP reform would create short-term political costs for incumbent ministers of agriculture. The myopic perspective of ministers of agriculture adversely affects the chance of any CAP reform being agreed by the CoAM; instead it creates a bias in favour of the status quo. Further, even when the short-term political calculations of the members of the CoAM allow a reform to be agreed, that reform will come from the limited set of reforms which do not upset the underlying political bargain in the CoAM. This bargain can be expressed in terms of each member state's net pay-off from the CAP. It is the baseline against which incumbent ministers of agriculture judge their performance. Hence no member will agree to any reform which adversely affects this.

The second step is to note that the CoAM is the lead institution on CAP matters. The public policy area of the EU is split into Councils of Ministers. These have specific functional responsibilities and capabilities to issue EU legislation in one division of the EU public policy area. Inevitably, there is a tendency for fiefdoms to emerge as each Council becomes protective of its turf in this public policy area. As noted in this work, these turfs can often overlap as certain policies create spillovers. The CAP is the largest item of expenditure from the EU budget and is the oldest common policy. Hence this tendency for a fiefdom to emerge for its control is particularly entrenched. Together these two steps form the insight of the institutions framework into why CAP reforms will only ever take the form of turbulence in the context of long-run stability.

The question still remains as to how the institutions framework can cope with the two main differences between the MacSharry reforms and the reforms of the CAP in 1984 and 1988. The first main difference is that the European Council was involved in the decision to reform the CAP in 1992 and that there was no real sense of crisis in the manner of 1984 and 1988. The institutions framework encourages consideration of the different pressures for CAP reform involved at different times. In the MacSharry reforms, there was no sense of crisis because the European Council was not involved. It was not involved because there was no overspill which necessitated the role of an arbitrator to mollify tension among the competing fiefdoms. There was no overspill with the budget issue or the international trade negotiation issue. As noted in Section 9.3, it was the toughness of the operation of the stabilizer regime which had an impact directly on the members of the CoAM.

9.4.1 The Role of Individuals

The second main difference noted in Section 9.3 was MacSharry and his personal ambitions for CAP reform. It was admitted in Chapter 5 that a possible weakness of the institutions framework was a lack of assessment and/or way of assessing the abilities and skills of the individual outside their institutional context; that is, MacSharry was an effective political operator during the reform process of 1990–1992. It was a factor in the start of the reform process as described under the heading of political leadership in Chapter 6.

The claim that individuals make a difference does not, to any substantial degree, reduce the usefulness of imposing a common framework on the various episodes of CAP reform being studied. This book has consistently understood the three frameworks of Chapter 5 as providing a loose description of the structure of the CAP decision-making system. As suggested by Przeworski (1985), frameworks can be used to define the context of public policy formation. The frameworks all set parameters for the actions of individual agents. The comparative study of the CAP reforms of 1984, 1988 and 1992 suggests that these parameters should be set reasonably wide. MacSharry used the position of Agriculture Commissioner in different ways and to greater effect than either Andriessen or Dolsager. However, they all shared a certain level of power and influence in the reform process simply because they were the Agriculture Commissioner. This common level of power and influence was sufficiently high in all three cases to make it worth imposing a common analytical framework which emphasizes the institutional structure as a key factor in the outcome of a reform process.

Individual political skills and influence should be understood in terms of their institutional context. In general, the Agriculture Commissioner is an important figure in a CAP reform process, but this importance can range from being very high (MacSharry) to merely moderate (Dolsager). These perhaps define the parameters of the power and influence of the position of the Agriculture Commissioner.

9.5 CONCLUDING REMARKS

The research conducted for the first objective of this book has supported the claim that the institutions framework is more insightful than the prominent players and interest groups frameworks in the analysis of CAP reforms. This claim fits in with one of the current issues in the political economy literature dealing with agricultural policies. A debate has emerged on the relative importance of interest groups compared with that of voters in formal models of the formation of agricultural policies in OECD countries. Both sides of the debate note that empirical research into the formation of agricultural polices requires attention to the factors contained within the institutions framework. Brooks (1995, p. 401) complains about the abstraction of many political economy models (citing the 'politician–voter' model of de Gorter and Swinnen (1994) as a prime example): 'their most telling weakness arises from the fact that they are inherently deterministic, discounting the importance of historical precedent, institutional structures, cultural values and political leadership'. In response to Brooks's general observation about the level of abstraction, de Gorter and Swinnen (1995, p. 413) state that 'we are in complete agreement with Brooks that we also need a model of institutions, constitutions, rules by which rules are made'. Hence both sides of the current debate admit that an understanding of policy outcomes requires an appreciation of past decisions and institutions and their interrelationships. This book provides such an appreciation with regard to the most significant event so far in the evolution of the agricultural policies of the EU: the 1992 MacSharry reforms.

Appendix: the Case Study Methodology

Petit *et al.* (1987) suggest a general approach to the development of a case study of the CAP policy process. It consists of four steps:

1. Identify the major participants in the policy process.
2. Describe their roles and behaviour and assess their influence.
3. Examine how a final compromise was reached in the CoAM.
4. Study how a given policy outcome affects the relative positions of various interests in the continuing policy debate.

This is the methodology adopted in conducting the fieldwork research for this book. The institutions framework suggests places or people to research in order to complete each of the four steps. As outlined in Chapter 5, the institutions framework provides ideas on who the major participants of the reform process are, what sorts of questions about their strategy and actions should be asked, and where the key parts of the reform process will take place.

The evidence for the case study has been gathered using two methods. The first was a critical review of the written material surrounding the MacSharry reforms: Commission documents, newspapers, academic journals and specialist agribusiness commentaries. The second method of research was a series of 17 in-depth interviews of participants in, or close observers of, the reform process. These were conducted in Brussels, London and Tunbridge Wells between October 1994 and March 1995. They were given on a non-attributional basis.

The *Financial Times*, *The Economist* and *The European* all gave a sound background to the reform process. However, they are aimed at non-agriculture or non-agricultural policy specialists. Their accounts tended to summarize the reform process as being neat and tidy (because it is easier to explain it that way). Further, the story of the reforms was told according to each of their 'world views' of government, business, the EU and the CAP. In contrast, *AgraEurope* provided much of the detail of the progress of the MacSharry reforms from their development in the MacSharry cabinet until their final enactment in May 1992. In particular, their correspondent,

Brian Gardner, always seemed to have an insider's insight and knowledge of the progress of the reform plans in both the Commission and the CoAM. Indeed, a number of the Commission officials I talked to mentioned Mr Gardner's reports as their own best source of information on how the reform process was developing. *Agence Europe*, although not focusing exclusively on the CAP, contained daily reports of the activities of the Commission and the CoAM over the entire reform process. These were valuable sources of information. Ross (1994, 1995) and Grant (1994), in their accounts of the Delors presidencies, provided evidence concerning the Delors–MacSharry relationship, which was at times important in the CAP reform process.

A list of the people interviewed is given in Table A.1, which gives their position at the time of the MacSharry reforms and the date of the interview. The interviews conducted can be described in terms of their structure, content and timing. Each interview lasted between 1 and 1.5 h, except as otherwise indicated. They were face-to-face and were systematic in the sense that there was some common set of questions in each interview. These questions were tightly specified so that the answers would be unambiguous. In advance of the interview, I prepared a series of topics and questions that I wished to cover. However, the interviews were left flexible, so that the interviewee could raise topics that he or she thought were important. Each of these off-hand remarks was followed up.

There is a particular type of evidence that an interview can provide that other secondary sources cannot. The public choice paradigm focuses on the individual. The intentional aspect of individual agent's behaviour is important in understanding the CAP reform process. Specifically, interviews provide details of the following areas:

- the distinct personalities of the important individuals in the reform process;
- those individuals' strategies, objectives and actions;
- those individuals' views of their own position and influence in the reform process;
- their view of other individuals in the reform process and other individuals' view of them;
- individuals' view of the role of institutions and their interrelationship in the outcome of the reform process.

The timing of the interviews, 2 years after the enactment of the reforms, was suited to academic research. Memories and recollections seemed relatively clear. The immediate consequences of the political confrontations in the reform process had been dissipated and most participants were open and objective. However, it is a general observation of the interviews that the closer the individual was to the reform process, the less objective was

Table A.1. People interviewed in connection with this book.

Name	Position	Date of interview
Raymond MacSharry	Agriculture Commissioner	5 October 1994[a]
Chris Horseman	Brussels Correspondent of *AgraEurope*	1 November 1994
Martin Haworth	Head of International Affairs, NFU	15 November 1994
Terry Wynn	MEP, Chairman of the LUFPIG and member of the EP's Budget Committee	23 November 1994
Peter Clinton	Rapporteur, EP's Agriculture Committee	23 November 1994
Jane Kelsey	Researcher, LUFPIG	23 November 1994
François Raynaud	Researcher, COPA	23 November 1994
Alan Wilkinson	DG VI	24 November 1994
John Slater	Current Head of Economics, MAFF	29 November 1994
Ron Irving	Previous Head of Economics, MAFF (during MacSharry reforms)	29 November 1994
David Frost	First Secretary, UKREP	14 December 1994
David Roberts	Deputy Director, DG VI	25 January 1995
John Gummer	UK Secretary of State for Agriculture	3 February 1995[b]
Justine Patterson	Council Secretariat for CoAM	16 February 1995
Phillip Rycroft	Currently member of the Brittan cabinet with overall responsibility for agriculture	6 March 1995
Guy Legras	Director, DG VI	6 March 1995[c]
Patrick Hennessy	Deputy Chef de Cabinet, MacSharry cabinet, with overall responsibility for agriculture	6 March 1995

[a] The interview with Raymond MacSharry was conducted over the telephone.
[b] The interview with John Gummer was conducted by post.
[c] The interview with Guy Legras was impromptu and lasted only 20 min.

his or her judgement of the process. The evidence of the interviews conducted supported the point from the public choice paradigm (see Chapter 5 for its outline) that an individual's perspective of the CAP and its reform is, to a large extent, conditioned by the institution to which that individual belongs and his or her position within that institution. The institutional context of individuals directly involved in the reform process emerged as a factor in that process's direction and overall outcome.

In addition, there was evidence of an important role for the individual within his or her institutional context. That is, personalities, specific experiences and manner were influential factors at certain points in the reform process. As noted in Chapter 4, this is something not specifically accounted for in the institutions framework. It is something that the case study for this book highlights as a material factor in a CAP reform process.

Bibliography

Ackrill, R.W. (1992) The EC budget and agricultural policy reforms, with special reference to cereals. Unpublished PhD thesis, University of Nottingham, Nottingham.

Ackrill, R.W., Hine, R.C., Rayner, A.J. and Suardi, M. (1994) The distributional effects of the Common Agricultural Policy between member states: budget and trade effects. CREDIT Research Paper, Department of Economics, University of Nottingham, Nottingham.

Ackrill, R.W., Hine, R.C., Rayner, A.J. and Suardi, M. (1995) The impact of the MacSharry reforms on the distributional effects of the Common Agricultural Policy between member states: budget and trade effects. CREDIT Research paper, Department of Economics, University of Nottingham, Nottingham.

AgraEurope (various issues).

Agence Europe (various issues).

Allinson, G. (1971) *Essence of Decision: Explaining the Cuban Missile Crisis.* Little, Brown, Boston, Massachusetts.

Alston, J.M. and Smith, V.H. (1994) If the CAP fits: political economy, policy inertia, export subsidies and CAP reform. Annual Conference Agricultural Economics Society, University of Exeter, April 1994.

Ardy, B. (1988) The national incidence of the European Community budget. *Journal of Common Market Studies* 26, 401–429.

Avery, G. (1984) Europe's agricultural policy: progress and reform. *International Affairs* 60, 643–656.

Averyt, W.F. (1977) *Agropolitics in the European Community: Interest Groups and the CAP.* Praeger, New York.

Bannister, P., Burman, E., Parker, I., Taylor, M. and Tindall, C. (1994) *Qualitative Methods in Psychology.* Open University Press, Buckingham.

Becker, G.S. (1976a) *The Economic Approach to Human Behaviour.* Cambridge University Press, Cambridge.

Becker, G.S. (1976b) Comment (on Peltzman). *Journal of Law and Economics* 19, 245–248.

Becker, G.S. (1983) A theory of competition among pressure groups for political influence. *Quarterly Journal of Economics* 98, 371–400.

Becker, G.S. (1985) Public policies, pressure groups, and deadweight costs. *Journal of Public Economics* 28, 329–347.

Beghin, J.C. (1990) A game theoretic model of endogenous public policies. *American Journal of Agricultural Economics* 72, 138–148.

Bentley, A.F. (1908) *The Process of Government.* University of Chicago Press, Chicago, Illinois.

Bernstein, R.J. (1976) *The Restructuring of Social and Political Theory.* Methuen, London.

Bhagwati, J.N. (1982) Directly unproductive profit-seeking activities. *Journal of Political Economy* 90, 988–1002.

Blaug, M. (1992) *The Methodology of Economics,* 2nd Edn. Cambridge University Press, Cambridge.

Bonner, J. (1986) *Politics, Economics and Welfare.* Wheatsheaf, Brighton.

Bowler, I.R. (1985) *Agriculture under the Common Agricultural Policy.* Manchester University Press, Manchester.

Bromley, D.W. (1989) *Economic Interests and Institutions.* Basil Blackwell, Oxford.

Brooks, J. (1995) The economic polity of farm policy: a comment. *Journal of Agricultural Economics* 46, 398–402.

Brooks, J. (1996) Agricultural policies in OECD countries: what can we learn from political economy models? *Journal of Agricultural Economics* 47, 366–389.

Brown, C. (1988) *Price Policies of the CAP: Retrospect and Prospect,* Report no. 41. Institute of Agricultural Economics, Copenhagen.

Buchanan, J.M. and Tullock, G. (1962) *The Calculus of Consent.* University of Michigan Press, Ann Arbor, Michigan.

Buckwell, A.E., Harvey, D.R., Thomson, K.J. and Parton, K.A. (1982) *The Costs of the Common Agricultural Policy.* Croom Helm, London.

Bulmer, S. and Wessels, W. (1987) *The European Council.* Macmillan, London.

Burton, M. (1985) The implementation of the EC milk quota. *European Review of Agricultural Economics* 12, 461–471.

Camps, M. (1967) *European Unification in the Sixties from the Veto to the Crisis.* Oxford University Press, Oxford.

Clout, H. (1984) *A Rural Policy for the EEC?* Methuen, London.

Cohen, R. (1959) *The Economics of Agriculture.* Cambridge University Press, Cambridge.

Commission of the European Communities (1983a) *Further Guidelines of the Development of the Common Agricultural Policy.* COM (83) 380. CEC, Brussels.

Commission of the European Communities (1983b) *Common Agricultural Policy: Proposals of the Commission.* COM (83) 500. CEC, Brussels.

Commission of the Europrean Communities (1987a) *Making a Success of the Single European Act: a New Frontier for Europe.* COM (87) 100. CEC, Brussels.

Commission of the European Communities (1987b) *On Financing the Community Budget.* COM (87) 101. CEC, Brussels.

Commission of the European Communities (1987c) *Review of Action Taken to Control the Agricultural Markets and Outlook for the Common Agricultural Policy.* COM (87) 410. CEC, Brussels.

Commission of the European Communities (1987c) *Own Resources Decision.* COM (87) 420. CEC, Brussels.

Commission of the European Communities (1987d) *On Budgetary Discipline.* COM (87) 430. CEC, Brussels.

Commission of the European Communities (1987e) *Implementation of Agricultural Stabilisers.* COM (87) 452. CEC, Brussels.

Commission of the European Communities (1991a) *The Development and Future of the Common Agricultural Policy*. COM (91) 100. CEC, Brussels.

Commission of the European Communities (1991b) *The Development and Future of the Common Agricultural Policy. Follow Up to the Reflections Paper COM (91) 100 1 February 1991*. COM (91) 258. CEC, Brussels.

Commission of the European Communities (1994) *EC Agricultural Policy for the 21st Century*. European Economy Reports and Studies no. 4. CEC, Brussels.

Commission of the European Communities (1996) *Directory of Interest Groups*. CEC, Brussels.

Cox, G., Lowe, P. and Winter, M. (1986) Agriculture and conservation in Britain: a policy community under siege. In: Cox, G., Lowe, P. and Winter, M. (eds) *Agriculture: People and Politics*. Allen and Unwin, London.

Cox, G., Lowe, P. and Winter, M. (1987) Farmers and the state: a crisis for corporatism. *Political Quarterly* 58, 73–81.

Cronbach, L.J. (1982) *Designing Evaluations of Educational and Social Programs*. Jossey-Bass, San Francisco.

Davey, B., Josling, T.E. and McFarquhar, A. (eds) (1976) *Agriculture and the State*. Macmillan, London.

Dinan, D. (1994) *Ever Closer Union? An Introduction to the European Community*. Macmillan, London.

Dixit, A. and Nalebuff, B. (1991) *Thinking Strategically*. Norton, New York.

Doering, O. (1991) Looking back while going forward: an essential for policy economists. *Journal of Agricultural Economics Research* 43, 3–6.

Downs, A. (1957) *An Economic Theory of Democracy*. Harper and Row, New York.

Dunleavy, P. (1991) *Democracy, Bureaucracy and Public Choice*. Harvester Wheatsheaf, London.

Eckstein, H. (1975) Case study and theory in political science. In: Greenstein, F.I. and Polsby, N.W. (eds) *Handbook of Political Science*, vol. 1: *Political Science: Scope and Theory*. Addison-Wesley, Reading, Massachusetts.

Edgell, J. and Thompson, K. (1997) NGOs and lobbying in the CAP. Paper presented to the Agricultural Economics Society Annual Conference, 1997.

Eichengreen, B. and Frieden, J. (1993) The political economy of European monetary unification: an analytical introduction. *Economics and Politics* 5, 85–104.

Elster, J. (1979) *Ulysses and the Sirens: Studies in Rationality and Irrationality*. Cambridge University Press, Cambridge.

Elster, J. (1983) *Explaining Technical Change: a Case Study in the Philosophy of Science*. Cambridge University Press, Cambridge.

Elster, J. (ed.) (1986) *Rational Choice*. Blackwell, Oxford.

Elster, J. (1989) *Nuts and Bolts for the Social Sciences*. Cambridge University Press, Cambridge.

Fearne, A.P. (1988) *Annual Price Review: a Framework for CAP Decision-Making*. Department of Agricultural Economics 2/88, University of Newcastle, Newcastle upon Tyne.

Fearne, A.P. (1989) A 'satisficing' model of CAP decision-making. *Journal of Agricultural Economics* 40, 71–81.

Fearne, A.P. (1991) The Council of Agricultural Ministers. In: Ritson, C. and Harvey, D. (eds) *The Common Agricultural Policy and the World Economy*. CAB International, Wallingford.

Fennell, R. (1987) *The Common Agricultural Policy of the European Community.* BSP Professional Books, Oxford.

Field, H. and Fulton, M. (1994) A bargaining model of the CAP. *American Journal of Agricultural Economics* 76, 15–25.

Field, H., Hearn, S. and Kirby, M.G. (1989) *The 1988 EC Budget and Production Stabilisers.* Discussion Paper, Australian Bureau of Agricultural and Resource Economics.

Frey, B.S. (1978) *Modern Political Economy.* Martin Robertson, Oxford.

Froud, J. and Roberts, D. (1993) The welfare effects of the new CAP cereals regime: a note. *Journal of Agricultural Economics* 44, 496–501.

George, S. (1991) *Politics and Policy in the European Community.* Oxford University Press, Oxford.

Gibbons, R. (1992) *A Primer in Game Theory.* Harvester Wheatsheaf, Hemel Hempstead.

Goetz, J.P. and LeCompte, M.D. (1984) *Ethnography and Qualitative Design in Educational Research.* Academic Press, Orlando, Florida.

Goldthorpe, J.H. (1984) *Order and Conflict in Contemporary Capitalism.* Oxford University Press, Oxford.

de Gorter, H. (1994) The Economic Polity of Farm Policy. Annual Conference, Agricultural Economics Society, University of Exeter, April.

de Gorter, H. and Swinnen, J. (1994) The economic polity of farm policy. *Journal of Agricultural Economics* 45, 312–326.

de Gorter, H. and Swinnen, J. (1995) The economic polity of farm policy: a reply. *Journal of Agricultural Economics* 46, 403–414.

de Gorter, H. and Tsur, Y. (1991) On the political economy of public goods inputs in agriculture. *American Journal of Agricultural Economics* 73, 1244–1254.

Grant, C. (1994) *The House that Jacques Built.* Nicholas Brearly, London.

Grant, W. (1989) *Pressure Groups, Politics and Democracy in Britain.* Philip Allan, New York.

Grant, W. (1993) Pressure groups and the European Community: an overview. In: Mazey, S. and Richardson, J.J. (eds) *Lobbying in the European Community.* Oxford University Press, Oxford.

Grant, W. (1995a) Is agricultural policy still exceptional? *Political Quarterly* 66, 156–169.

Grant, W. (1995b) The limits of CAP reform and the option of denationalisation. *Journal of European Public Policy* 1, 1–18.

Green, D.P. and Shapiro, I. (1994) *Pathologies of Rational Choice Theory.* Yale University Press, New Haven, Connecticut.

Hagedorn, K. (1983) Reflections on the methodology of agricultural policy research. *European Review of Agricultural Economics* 10, 303–323.

Hagedorn, K. (1985) CAP reform and agricultural economics: a dialogue of the deaf? In: Pelkmans, J. (ed.) *Can the CAP be reformed?* European Institute of Public Administration, Maastricht.

Hallett, G. (1968) *The Economics of Agricultural Policy.* Basil Blackwell, Oxford.

Hammersley, M. (1989) *The Dilemma of Qualitative Method: Herbert Blumer and the Chicago Tradition.* Routledge, London.

Hargreaves-Heap, S., Hollis, M., Lyons, B., Sugden, R. and Weale, A. (1992) *The Theory of Choice.* Basil Blackwell, Oxford.

Harvey, D. (1982) National Interests and the CAP. *Food Policy*, 1, 76–90.

Hathaway, D. (1963) *Government and Agriculture.* Macmillan, London.

Hempel, C.G. (1965) *Aspects of Scientific Explanation.* Free Press, New York.

Henwood, K.L. and Pidgeon, N.F. (1993) Qualitative research and psychological theorizing. In: Hammersley, M. (ed.) *Social Research Philosophy, Politics and Practice.* Open University Press, London.

Hine, R.C. (1973) Structural policies and British agriculture. *Journal of Agricultural Economics* 24, 321–329.

Hirschman, A.O. (1970) *Exit, Voice and Loyalty: Responses to Declines in Firms, Organizations and States.* Harvard University Press, Cambridge, Massachusetts.

Howarth, R.W. (1985) *Farming for the Farmers? A Critique of Agricultural Support Policy.* IEA, London.

Hubbard, L. and Ritson, C. (1991) The reform of the CAP. In: Ritson, C. and Harvey, D. (eds) *The Common Agricultural Policy and the World Economy.* CAB International, Wallingford.

Hull, R. (1993) Lobbying Brussels: a view from within. In: Mazey, S. and Richardson, J. (eds) *Lobbying in the European Community.* Oxford University Press, Oxford.

Ingersent, K.A. and Hill, B.E. (1982) *An Economic Analysis of Agriculture.* Heinemann, London.

Ingersent, K.A., Rayner, A.J. and Hine, R.C. (1993) Agriculture and GATT. In: Raynor, A.J. and Coleman, D. (eds) *Current Issues in Agricultural Economics.* Macmillan, London.

Ingersent, K.A., Rayner, A.J. and Hine, R.C. (1995) The Uruguay Round Agriculture Agreement. *The World Economy* 18, 707–728.

Johnson, D.G. (1973) *World Agriculture in Disarray.* Fontana-Collins, London.

Johnson, G.L. (1960) The labour utilisation problem in European and American agriculture. *Journal of Agricultural Economics* 14, 325–349.

Jordan, A.G., Maloney, W.A. and McLaughlin, A.M. (1994) Characterizing agricultural policy-making. *Public Administration* 72, 505–526.

Josling, T.E. (1973) The reform of the Common Agricultural Policy. In: Evans, D. (ed.) *Britain in the EC.* Gollancz, London, pp. 86–99.

Josling, T.E. (1974) Agricultural policies in developed countries: a review. *Journal of Agricultural Economics* 25, 220–264.

Keohane, R. and Hoffman, S. (eds) (1991) *The New European Community.* Westview Press, Boulder, Colorado.

King, G. (1989) *Unifying Political Methodology: the Likelihood of Statistical Inference.* Cambridge University Press, New York.

King, G., Keohane, R.O. and Verba, S. (1994) *Designing Social Inquiry.* Princeton University Press, Princeton, New Jersey.

Kjeldahl, R. and Tracy, M. (eds) (1994) *Renationalisation of the Common Agricultural Policy?* Institute of Agricultural Economics, Copenhagen, and Agricultural Policy Studies, Brussels: Combined Book Services.

Koester, U. (1978) Decision-making problems in the Council of Agriculture Ministers. *Intereconomics* 9, 211–215.

Koester, U. and Tangermann, S. (1977) Supplementary farm price policy by direct income payments. *European Review of Agricultural Economics* 13, 7–31.

Krueger, A.O. (1974) The political economy of rent-seeking society. *American Economic Review* 64, 291–303.

Kuhn, T.S. (1962) *The Structure of Scientific Revolutions.* University of Chicago Press, Chicago, Illinois.

Larsen, A. (1993) *Danish Agricultural Economy.* Reprinted in Commission of the European Communities (1994). *EC Agricultural Policy for the 21st Century.* European Economy Reports and Studies no. 4.

Lijphart, A. (1971) Comparative politics and comparative method. *American Political Science Review* 65,682–698.

Lucas, W. (1974) *The Case Survey Method: Aggregating Case Experience.* Rand, Santa Monica, California.

McCormick, R.E. and Tollinson, R.D. (1981) *Politicians, Legislation, and the Economy: An Inquiry into the Interest-Group Theory of Government.* Martinus Nijhoff, Boston, Massachusetts.

McCrone, G. (1962) *The Economics of Subsidising Agriculture.* Allen and Unwin, London.

Magee, S.P., Brock, W.A. and Young, L. (1989) *Black Hole Tariffs and Endogenous Policy Theory: Political Economy in General Equilibrium.* Cambridge University Press. Cambridge.

Marsh, D. and Rhodes, R.A.W. (1992) *Policy Networks in British Government.* Oxford University Press, Oxford.

Marsh, D. and Stoker, G. (1995) *Theory and Methods in Political Science.* Macmillan, London.

Marsh, J.S. (1977) Europe's agriculture: reform of the CAP. *International Affairs* 53, 604–614.

Marsh, J.S. (1985) Can budgetary pressures bring CAP reform? In: Pelkmans, J. (ed.) *Can the CAP Be Reformed?* Europe Institute of Public Administration, Maastricht.

Marsh, J.S. (1987) Alternative policies for agriculture in Europe. *European Review of Agricultural Economics* 14, 5–20.

Marsh, J.S. and Ritson, C. (1971) *Agricultural Policy and the Common Market.* Chatham House PEP, London.

Marsh, J.S. and Swanney, P.J. (1980) *Agriculture and the European Community.* Allen and Unwin, London.

Marsh, J.S. and Tangermann, S. (1996) Preparing Europe's rural economy for the 21st century. Report to the Land Use and Food Policy Inter-Group (LUFPIG) in the European Parliament.

Martin, A. (1958) *Economics and Agriculture.* Routledge and Kegan Paul, London.

Mazey, S. and Richardson, J.J. (eds) (1993) *Lobbying in the European Community.* Oxford University Press, Oxford.

Meester, G. (1987) Budgetary constraints and the international realities of the CAP. *European Review of Agricultural Economics* 14, 37–48.

Meester, G. and Van der Zee, F. (1993) EC decision-making, institutions and the CAP. *European Review of Agricultural Economics* 20, 131–151.

Miliband, R. (1969) *The State in Capitalist Society.* Basic Books, New York.

Milward, A. (1992) *The European Rescue of the Nation-state.* Routledge, London.

Ministry of Agriculture, Fisheries and Food (1995) *European Agriculture: the Case for Radical Reform.* MAFF, London.

Morris, N. (1980) The Common Agricultural Policy. *Fiscal Studies* 11, 17–35.

Moyer, H.W. and Josling, T.E. (1990) *Agricultural Policy Reform: Politics and Process in the EC and USA.* Harvester Wheatsheaf, Hemel Hempstead.

Mueller, D.C. (1989) *Public Choice II.* Cambridge University Press, Cambridge.

Nardone, G. and Lopez, R.A. (1994) The welfare effects of the new CAP cereals regime: a comment. *Journal of Agricultural Economics* 45, 386–388.

National Farmers' Union (1994) *The GATT Settlement in Agriculture.* NFU

Nedergaard, P. (1994) The political economy of CAP reform. In: Kjeldahl, R. and Tracy, M. (eds) *Renationalisation of the Common Agricultural Policy?* Institute of Agricultural Economics, Copenhagen, and Agricultural Policy Studies, Belgium: Combined Book Services.

Neville-Rolfe, E. (1984) *The Politics of Agriculture in the European Community.* European Centre for Policy Studies, London.

Niskanen, W. (1971) *Bureaucracy and Representative Government.* Aldine-Atherton, Chicago, Illinois.

Niskanen, W. (1973) *Bureaucracy, Servant or Master?* Institute of Economic Affairs, London.

Northglass, D. (1988) Public choice models. In: Elster J. (ed.) *Rational Choice.* Blackwell, Oxford.

Ockenden, J. and Franklin, J. (1995) *European Agriculture: Making the CAP Fit the Future.* Chatham House, London.

Olson, M. (1965) *The Logic of Collective Action.* Harvard University Press, Cambridge, Massachusetts.

Olson, M. (1977) *The Logic of Collective Action: Public Goods and the Theory of Groups.* Harvard University Press, Cambridge, Massachusetts.

Olson, M. (1985) Space, agriculture and organization. *American Journal of Agricultural Economics* 67, 928–937.

Olson, M. (1986) The exploitation and subsidization of agriculture in developing and developed countries. In: Maunder, A. and Renborg, U. (eds) *Agriculture in a Turbulent World Economy.* Gower, Aldershot, pp. 49–59.

Olson, M. (1990) *Agricultural Exploitation and Subsidization: There Is an Explanation.* Gower, Aldershot.

Pearce, J. (1981) *The Common Agricultural Policy.* Routledge and Kegan Paul, London.

Pelkmans, J. (ed.) (1985) *Can the CAP Be Reformed?* European Institute of Public Administration, Maastricht.

Peltzman, S. (1976) Toward a more general theory of regulation. *Journal of Law and Economics* 19, 211–240.

Peterson, J. (1995) Decision-making in the European Union: towards a framework for analysis. *Journal of European Public Policy* 1, 69–95.

Petit, M., De Bendictis, M., Britton, D., De Groot, M., Henrichsmeyer, W. and Lechi, F. (1987) *Agricultural Policy Formation in the European Community: the Birth of Milk Quotas and CAP Reform.* Elsevier, Amsterdam.

Posner, R.A. (1974) Theories of economic regulation. *Bell Journal of Economics and Management Science* 29, 335–357.

Posner, R.A. (1975) The social costs of monopoly and regulation. *Journal of Political Economy* 83, 807–827.

Przeworski, A. (1985) *The State and the Economy under Capitalism.* Cambridge University Press, Cambridge.

Putnam, R.D. (1988) Diplomacy and domestic politics: the logic of two-level games. *International Organization* 42, 428–460.

Ramusen, E. (1989) *Games and Information.* Basil Blackwell, Oxford.

Rausser, G.C. (1982) Political economic markets: PERTs and PESTs in food and agriculture. *American Journal of Agricultural Economics* 64, 821–833.

Rausser, G.C. (1989) The political economy of agricultural policy reform. *European Review of Agricultural Economics* 16, 349–366.

Rausser, G.C. and Freebairn, J.W. (1974) Estimation of policy preference functions: an application to the US beef import quotas. *Review of Economics and Statistics* 56, 437–449.

Rayner, A.J. and Colman, D. (eds) (1993) *Current Issues in Agricultural Economics.* Macmillan, London.

Rayner, A.J., Ingersent, K.A. and Hine, R.C. (1993a) Agriculture in the Uruguay Round: an assessment. *Economic Journal* 104, 1513–1528.

Rayner, A.J., Ingersent, K.A., Hine, R.C. and Ackrill, R.W. (1993b) *Does the CAP Fit the GATT? A Model of EU Cereals.* CREDIT paper 93/14, University of Nottingham, Nottingham.

Richardson, J.J. (ed.) (1982) *Policy Styles in Western Europe.* Allen and Unwin, London.

Richardson, J.J. and Jordan, A.G. (1979) *Governing under Pressure: the Policy Process in a Post-parliamentary Democracy.* Martin Robertson, Oxford.

Riker, W. (1982) *Liberalism Against Populism.* W.H. Freeman, San Francisco.

Ritson, C. (1977) *Agricultural Economics and Policy.* Granada, London.

Rose, R. (1991) *Comparing Forms of Comparative Analysis.* Studies in Public Policy no. 188. Centre of Studies of Public Policy, University of Strathclyde.

Ross, G. (1994) Inside the Delors Cabinet. *Journal of Common Market Studies* 32, 499–525.

Ross, G. (1995) *Jacques Delors and European Integration.* Polity Press, Cambridge.

Rowley, C.K., Tollison, R.D. and Tullock, G. (eds) (1988) *The Political Economy of Rent-Seeking.* Kluwer Academic, Boston, Massachusetts.

Runge, C.F. and Von Witzke, H. (1987) Institutional changes in the CAP of the EC. *American Journal of Agricultural Economics* 69, 213–223.

Samuels, W.J. (1979) *The Economy as a System of Power.* Transaction Books, New Brunswick, New Jersey.

Sartori, G. (1976) *Parties and Party Systems: a Framework for Analysis,* Vol. 1. Cambridge University Press, Cambridge.

Schmitt, G. (1986) Agricultural policy decisions in the EC. *Food Policy* 5, 334–345.

Self, P. and Storing, H. (1962) *The State and the Farmer.* Allen and Unwin, London.

Senior-Nello, S. (1984) An application of public choice theory to the question of CAP reform. *European Review of Agricultural Economics* 11, 261–283.

Senior-Nello, S. (1989) European interest groups and the CAP. *Food Policy* 2, 101–106.

Smith, M.J. (1990) *The Politics of Agricultural Support in Britain.* Dartmouth, Aldershot.

Smith, M.J. (1993) *Pressure, Power and Society.* Harvester Wheatsheaf, Hemel Hempstead.

Sodersten, B. and Reed, G. (1994) *International Economics,* 3rd Edn. Macmillan, London.

Stigler, G.J. (1971) The theory of economic regulation. *Bell Journal of Economics and Management Science* 2, 3–21.

Stigler, G.J. (1975) *The Citizen and the State.* University of Chicago Press, Chicago, Illinois.

Swinbank, A. (1989) The CAP and the politics of European decision making. *Journal of Common Market Studies* 2, 303–322.

Swinbank, A. (1993) CAP reform, 1992. *Journal of Common Market Studies* 3, 359–371.

Swinbank, A. (1994) EU agricultural trade relations with its neighbours. Paper presented to CREDIT Conference, University of Nottingham, April.

Swinnen, J. and Van der Zee, F.A. (1993) The political economy of agricultural policies: a survey. *European Review of Agricultural Economics* 20, 261–291.

Tangermann, S. (1992) *CAP Reform? In for a Penny, In for a Pound.* IEA Inquiry Paper.

Tangermann, S. (1996) CAP Reform: What Next? An ex-post review of the 1992 MacSharry reform. Paper presented to CREDIT conference, University of Nottingham, Nottingham.

Tracy, M. (1989) *Government and Agriculture in Western Europe 1880–1988*, 3rd Edn. Granada, London.

Tracy, M. (1994) In the spirit of Stresa. *European Review of Agricultural Economics* 21, 357–374.

Tullock, G. (1965) *The Politics of Bureaucracy.* Public Affairs Press, Washington, DC.

Tweeten, L. (1971) *Foundations of Farm Policy.* University of Nebraska Press, Lincoln, Nebraska.

Von Cramon-Taubadel, S. (1993) The reform of the CAP from a German perspective. *Journal of Agricultural Economics* 44, 394–409.

Von Witzke, H. (1986) Endogenous supranational policy decisions: the Common Agricultural Policy of the European Community. *Public Choice* 48, 157–174.

Wallace, H. and Wallace, W. (1996) *Policy-making in the European Union.* Oxford University Press, Oxford.

Ward, H. and Edwards, G. (1990) Chicken and technology: the politics of the European Community's budget for research and development. *Review of International Studies* 16, 111–129.

Weigall, D. and Stirk, P. (1992) *The Origins and Development of the European Community.* Leicester University Press, Leicester.

Wilkinson, A. (1994) The renationalisation of the CAP. In: Kjeldahl, R. and Tracy, M. (eds) *Renationalisation of the Common Agricultural Policy?* Institute of Agricultural Economics, Copenhagen, and Agricultural Policy Studies, Brussels.

Winters, L.A. (1987) The political economy of the agricultural policy of the industrial countries. *European Review of Agricultural Economics* 14, 285–305.

Winters, L.A. (1990) The so-called 'non-economic' objectives of agricultural support. In: *Modelling the Effects of Agricultural Policies.* OECD Economic Studies no. 13, OECD, Paris.

Yin, R.K. (1984) *Case Study Research: Design and Methods.* Sage, Beverley Hills, California.

Yin, R.K. and Heald, K.A. (1975) Using the case survey method to analyze policy studies. *Administrative Science Quarterly* 20, 371–381.

Index